Contents

Introduction   5

1   Collecting, handling and storing micro-organisms   6
     Simple sources: prepared foods, hay infusions, rain water, Winogradsky columns, soil   6
     Sterilisation: autoclaves, sterilising ovens, flaming, filtration, ultraviolet radiation, steam sterilisation, chemical sterilisation or disinfection   8
     Culture media   9
     More sources: air, water, soil   11
     Inoculating   12
     Safety precautions   13
     Manipulative procedures   13
     Keeping stock cultures   14
       Fungi and bacteria: slopes, stab cultures, broth cultures, deep freeze, freeze dried cultures, silica gel dried cultures   14
       Algae and most protozoa; biphasic cultures   15
       Free-living amoebae   16

2   Working with bacteria   17
     Some methods for staining bacteria: preparation of bacterial smears, methylene blue, gram staining, wet mounts   17
     Destroying bacteria: physical treatments—autoclave, ultraviolet light. Chemical agents—disinfectants, antibiotics   19
     Personal hygiene: finger prints, toilet training   22
     Food bacteriology: preservation of foods—heat applied to milk, use of inhibitory materials   23
     Quality control: methylene blue test for milk, resazurin test for milk, *Clostridium perfringens* in milk, presumptive coliform test   24
     Yoghurt production—a food fermentation process   27
     Microbial spoilage: protein breakdown, lipolysis, pectolysis, starch breakdown   28
     Spoilage of canned food   28
     Spoilage of pre-cooked foods   29
     Selective culture of bacteria: azotobacter, aerobic bacteria, denitrifying bacteria, desulphovibrio, *Clostridium pasteurianum*   29
     Estimating numbers of microbes: viable counts, total counts   32
     Optical density measurements   35
     Determination of cell dry weight   37
     The bacterial growth curve   37

Diauxic growth   38
Teaching method   39

3   Mycological investigations   40
     Simple methods for demonstrating the presence of fungi in the air: spores in the air, spores from decaying leaves, from mouldy substrates, isolation from soil   40
     General characteristics of organisms which may develop on exposed plates and those treated with gros inocula: mucor, rhizopus, absidia, aspergillus, penicillium, cladosporium, alternaria, candida, botrytis, trichoderma   41
     Growing fungi in culture: preparation of agar, preparation of slopes, preparation of plates, inoculation of slopes or plates from agar culture   42
     Handling fungal cultures, especially dry spored species   43
     Methods of examining cultures: slide cultures, culture on media with low concentration of nutrients   44
     Preparation of slides: method of mounting, staining, permanent slides   44
     Some investigations with plate cultures: growth of fungi on solid media, determination of the optimum temperature for fungal growth, mutation in *Aspergillus nidulans*—mutagenesis, nutritional mutants   45
     Growth of fungi in liquid cultures: an investigation of the growth of *Aspergillus* in liquid cultures of differing compositions, other growth investigations with *Aspergillus* in liquid culture   48
     The kinetics of some biochemical processes taking place around yeast cells   49
     Fungal pathogens of plants: principles and method for investigating the causal agents responsible for fungal rots, some questions which may be used to establish infectious conditions, host/pathogen combinations suitable for class investigation   50
     Investigating damping-off fungi: raising test seedlings, inoculating seedlings, isolation of fungi from damped-off seedlings, methods for eliminating bacterial contamination   51
     Rotting fungi   53
     Teaching potential   53

4 Little plants and animals   54

A demonstration of population growth:
preparation of population, examining the
data   54

Investigating the growth requirements of two
algae: preparing nutrients, an investigation
using *Hydrodictyon*, an investigation using
Chlorella   56

Production of growth inhibitors by algae: an
antibiotic from *Chlorella*, antibiotics from
larger algae   57

Preparing a pure strain culture by cleaning:
methods, an example of possible
procedures   58

Maintaining axenic cultures   59

Making cotton wool plugs   60

The isolation and culture of small amoebae   60

Teaching method   60

Appendix 1   61

Sources of information, publications,
sources of apparatus, chemicals and
living material

Appendix 2   63

Bibliography and references

A note on units[19]

Autoclave pressures and temperatures.

One pound force per square inch ($1 \, \mathrm{lbf/in^2}$)
equals 6.89476 kilo Newtons per square metre
($6.89476 \, \mathrm{kN/m^2}$).

One Newton per square metre ($\mathrm{N/m^2}$) equals
one Pascal (Pa).

In this book $1 \, \mathrm{lbf/in^2}$ has been equated to
$6.9 \, \mathrm{kN/m^2}$. Thus $10 \, \mathrm{lbf/in^2} = 69 \, \mathrm{kN/m^2}$, and
$15 \, \mathrm{lbf/in^2} = 103.5 \, \mathrm{kN/m^2}$.

$69 \, \mathrm{kN/m^2}$ pressure in an autoclave is
approximately equivalent to a temperature of
$115^\circ\mathrm{C}$; $103.5 \, \mathrm{kN/m^2}$ to $121^\circ\mathrm{C}$.

Volume capacities of bottles.

One fluid ounce (oz) equals 28.4131 cubic
centimetres ($\mathrm{cm^3}$)

In this book 1 oz has been equated to $28.4 \, \mathrm{cm^3}$
and the volumes of bottles have been rounded
to the nearest whole number.

The litre.

One litre equals one cubic decimetre ($\mathrm{dm^3}$).

In this book the term litre has been used
throughout.

Length.

One Angstrom (Å) equals $10^{-10} \, \mathrm{m}$.

Media.

Details of the preparation of nutrient agar are
found in Chapter 1, page 11 and malt agar in
Chapter 3, page 42.

The majority of the investigations described are suitable for use at the secondary stage of education with pupils
following courses of the type leading to the C.S.E. and G.C.E. 'O' level examinations. Some of them, however,
are more complex and if they are to be undertaken successfully, previous experience of the techniques
involved is essential. These are marked [AL] following the title and are suitable for work in Advanced level
biological science courses of the G.C.E. and beyond.

Schools Council
Educational Use of Living Organisms
General Editor: P J Kelly

Micro-organisms

Author: Peter Fry

Contributors: B Bainbridge
              G Holt

HODDER AND STOUGHTON
LONDON SYDNEY AUCKLAND TORONTO

Schools Council Educational Use of Living Organisms
Project
Director: P J Kelly
Research Fellow: J D Wray
This project was established at the Centre for Science
Education, Chelsea College, in 1969. Its main aims
have been to determine the needs of schools with
respect to living organisms, to evaluate the usefulness
of various kinds of organisms for educational
purposes, and to devise maintenance techniques and
teaching procedures for the effective use of
appropriate species.

The project was initiated by the Institute of
Biology and Royal Society Biological Education
Committee and received its major financial support
from the Schools Council. The Nuffield Foundation
and Harris Biological Supplies also gave generous
contributions.

Acknowledgements

Thanks are due to the following for the
provision of photographs: J D Wray (Figure 16b); Dr
R Davenport (Figure 18); Philip Harris Biological
Limited (Figure 21).
Figures 2, 3, 4, 5, 7, 8, 11, 12, 14, 16a, 26, 27,
and 29 were provided by the author.

ISBN 0 340 17052 2

First published 1977

Printed and bound in Great Britain for
Hodder and Stoughton Educational,
a division of Hodder and Stoughton Ltd,
Mill Road, Dunton Green, Sevenoaks, Kent.
by Unwin Brothers Limited
Computer Typesetting by Print Origination,
Merseyside, L20 6NS

Preface

The use of living organisms in schools, while not new,
has received considerable emphasis in recent develop-
ments in the teaching of biology, environmental
studies and allied subjects in colleges and secondary
schools, and in many aspects of primary school work.
It is a change reflecting the much wider movement
in educational thinking which acknowledges both the
interest and delight that young people can gain from
living things and the importance of fostering an
appreciation of the scientific, social, aesthetic and
moral issues involved in the study of life and the
natural environment. It is a change, also, that presents
some very real—but not insurmountable—practical
problems for schools.

The books in the Educational Use of Living
Organisms series deal with the principles which
underlie the effective educational use of living things.
They also provide information to help teachers
integrate work with organisms into their courses and
to cope with the practical day-to-day problems
involved. For teachers and administrators there are
technical details of value for planning facilities, and
annotated bibliographies provide the guidelines for
more detailed studies if required.

Microbiology is featured in many courses and its
educational importance is clearly recognised. In this
book Peter Fry and his colleagues have provided
educational and technical guidance of considerable
value to those who wish to undertake practical work
using micro-organisms. From their own work and
through their membership of the Microbiology in
Schools Advisory Committee, they have been able to
call on a wide range of experience and advice. They
have also been helped by Mr. J. D. Wray who undertook
much of the research of the EULO project.

All the investigations as described in this book
have been carried out safely and without accident in
schools. Nevertheless, it should be stressed that the
same level of care should be taken in using micro-
organisms as one would exercise in using potentially
harmful chemicals in a school laboratory. Micro-organisms
(especially bacteria) can be dangerous if handled
carelessly.

Teachers contemplating work with micro-organisms
are advised to have some training in the techniques
involved. Indeed, I hope that the publication of this
book will encourage the establishment of more short
courses in practical microbiology for teachers.

P. J. Kelly

Introduction

Micro-organisms are everywhere: in the air, in the soil and in the water of the Earth and, because of their ubiquitous distribution, they are important not only in the overall economy of Man but also as organisms which illustrate major biological principles for education. Through their metabolic activities a wide range of chemicals essential to life pass from one ecosystem to another. Soil bacteria, and bacteria in symbiotic relationships with leguminous plant roots, take up nitrogen from the air; others return the same gas as 'a result of feeding upon nitrogenous substrates. Deciduous trees, casting their leaves in autumn, are in many ways the original 'litter louts' depositing their sheets of leaf cellulose on the ground beneath their branches as man, in our developed society, leaves his paper wrappings. But fortunately the material does not stay on the ground for long; a veritable host of micro-organisms produce cellulases which break down the cellulose and feed upon the rich substrate. Equally, of course, many micro-organisms feed upon the rich substrate of the bodies of other living plants and animals. Devoted gardeners fight desperately to avoid mildew infections on their roses and everybody would prefer to avoid being host to the rich fauna of pathogenic bacteria and viruses.

In using micro-organisms for biology teaching it is interesting to note how our self-concern so frequently biases our choices. We often start work with bacteria as examples of 'germs' and whilst no introductory course in microbiology would be complete without some consideration of personal hygiene perhaps consideration of the overall affects of fungi in biological communities might give students a better appreciation of the place of this fascinating group in the natural economy of life. And, at the same time, we can then lead them towards an understanding of the application of biological principles to satisfying the needs of man, the area of activity we normally describe as applied biology. Prior to 1923 at least 90% of the world's citric acid was derived annually from the Italian fruit harvest. In that year a factory was opened in New York in which this important culinary acid, important as an organic compound itself and as a starting point for the syntheses of others, was made by 'fermentation' of sucrose using *Aspergillus niger*. At a stroke the cost of citric acid dropped to one tenth of the original price. Another species of the same genus, *Aspergillus ochraceus*, is used in the preparation of the drug cortisone and a wide range of other fungal and bacterial species are important sources of chemical and food materials.

The area of activity which teachers term 'education' has many similarities to the 'good loam' of gardeners. It means different things to different people and no one definition out of context can be said to be in any way superior to another. Only the individual teacher can select the 'right' examples to illustrate fundamental biological ideas to his students. This book is thus a collection of procedures which can be used in a variety of ways to present and extend biological concepts using micro-organisms. It is a source of ideas, but not a course in microbiology in any sense.

Details of the use of micro-organisms for investigations in genetics can be found in the companion publication 'Organisms for Genetics'[10] and a wide ranging discussion of the potential uses of living organisms can be found in Appendix 2, reference 11.

Sources of information and of living material, associated equipment and apparatus will be found in Appendix 1.

When working with micro-organisms both student and teacher will inevitably have some surprises. These are both stimulating and fascinating as starting points for research in the truest sense of the word. How much time is available actually to follow up problems which arise depends upon the nature of the course which students are taking, but there can be no doubt that every young scientist should discover some of the magic of the scientific discipline as well as appreciate some of the facts of science.

It would be foolish to suggest that any idea in a book of this kind is truly original. Each has been culled from many sources and the corporate experience of biology teachers who have attended microbiology courses in the London area over the last decade has been a particularly rich resource.

Chapter 1 Collecting, handling and storing micro-organisms

Practical work in microbiology can be done with comparative ease and economy. It is not essential to buy expensive cultures as there are rich sources of naturally occurring micro-organisms which, if handled carefully, do not give rise to health hazards.

The basic techniques for handling micro-organisms were established many years ago, mainly with regard to bacteria, but although considerable technical advances have been, and are still being made, these techniques enable workers to handle the organisms with efficiency. They prevent the escape of organisms from containers and they prevent unwanted organisms entering an experimental set-up so interfering with the work in hand.

In order to grow, micro-organisms must be provided with the food materials they need. Aqueous plant or meat extracts provide ideal substrates or *media*. They may be used as liquids (*broths*) or they may be solidified by the addition of the inert carbohydrate, agar, which is derived from sea-weed, like *Gracilaria confervoides* of the shores of Malaysia, Indonesia and Japan. This produces a hydrophilic colloid whose gel form is strong so that it does not fall out of up-turned containers nor does its surface score when streaked. It does not become a sol until heated to $98°C$ and yet it remains as a sol until it is cooled to $35-40°C$. This is low enough for most organisms to be added without the organisms being heat killed.

SIMPLE SOURCES
Some simple natural sources of micro-organisms are:

1 Prepared foods
Most cooked foods will go 'off' if left in warm, damp surroundings. The ideal materials for this are foods with solid surfaces so that colony growth can be observed, e.g. cottage cheese, bread slices and sliced vegetables such as potato, carrot and turnip. Homemade bread is best as this goes mouldy more easily due to the lack of the chemical preservatives normally added to the commercially produced loaf. The foods should be placed on damp paper towelling

in containers like pudding basins and covered with polythene or foil and finally placed in a warm dark place for several days.

In all of these examples, bacteria, yeasts and moulds can be examined directly under the microscope and students can build up some quantitative information about the numbers and variety of micro-organisms which do exist around us in ordinary places like the domestic kitchen.

2 Hay infusions
Boil about 1g of chopped hay or dried grass cuttings with approximately 100 cm^3 of pond or rain water and allow the resulting medium to cool. Next add a little freshly chopped hay and allow the whole mix to stand in a glass beaker with a paper cover. After some days a surface scum of bacteria will develop. This should not be broken and in subsequent weeks algae and protozoans like *Colpidium* and *Paramecium* will become established. In some regions of the country better results can be obtained if a couple of 'marble chips' are added to the brew.

3 Rain water
Stand a clean, folded filter paper in a funnel outside during a rain storm. Ideally rain water should run through the paper for about a half-day. Next open the paper out and lay it flat on a plate, 'feed' it with a little very dilute house plant nutrient (like 'Baby Bio') and cover it with a transparent cover. Keep the paper damp with chlorine and copper free water or just rain water and observe it after some days when algae and fungal colonies have usually become established.

4 Winogradsky columns
Fill graduated cylinders or similar tall containers with suitable mud and water to simulate pond conditions. When left for several months, a succession of algae and sulphur bacteria are induced. This simple demonstration illustrates the principle of enrichment culture and because of the pigmented nature of most sulphur metabolising bacteria their succession is clearly seen.

Equipment and chemicals
  500 cm³ glass measuring cylinder
  Shredded filter paper, say 6–8 papers
  River or pond mud, about 150 cm³
  Calcium sulphate, 20g ($CaSO_4 . \frac{1}{2}H_2O$ 'Plaster of
    Paris)
  Solution of ammonium chloride (NH_4Cl) in
    sodium phosphate buffer (0.02M, pH 7.4),
    about 300 cm³ will be needed
The phosphate buffer is made from solutions A and
B mixed together in the proportions of 19 : 81
  Solution A : 0.02M sodium dihydrogen
    orthophosphate ($NaH_2PO_4 . 2H_2O$) 2.8g in 1
    litre of water
  Solution B : 0.02M disodium hydrogen
    orthophosphate ($Na_2HPO_4 . 7H_2O$) 5.4g in 1
    litre of water
The buffered ammonium chloride solution is made by
dissolving 0.1g of ammonium chloride to every
100 cm³ phosphate buffer solution.

Setting up
  Mix the mud with the calcium sulphate and the
shredded filter papers and transfer the slurry to the
measuring cylinder. Overlay the mud with the
buffered ammonium chloride solution, making sure
that the sides of the cylinder above the mud-water
interface are clear of mud. Leave the soil column on
a well illuminated window sill and observe the
changes that occur.
  Sulphate-reducing bacteria in the mud generate
hydrogen sulphide and conditions become
increasingly anaerobic as a result of this process and
the action of fermentative aerobes which utilise
the cellulose. Green algae frequently develop at the
surface of the liquid phase and generate oxygen; they
may disappear as the hydrogen sulphide
concentration rises to toxic levels. With illumination,
colonies of purple (*Chromatium*) and green
(*Chlorobium*) photosynthetic bacteria develop at the
glass-mud interface at the expense of the hydrogen
sulphide produced by bacteria such as
Desulphovibrio. Sulphate is regenerated by the
activity of non-pigmented bacteria (*Thiobacillus*);
the distribution of the latter in the liquid phase will
be controlled by the respective concentrations of
oxygen and hydrogen sulphide.
  The black, sulphate reducers should begin to
appear within one week and the development of
pigmented, photosynthetic bacteria may be
anticipated some two to three weeks later (see Fig.
1).

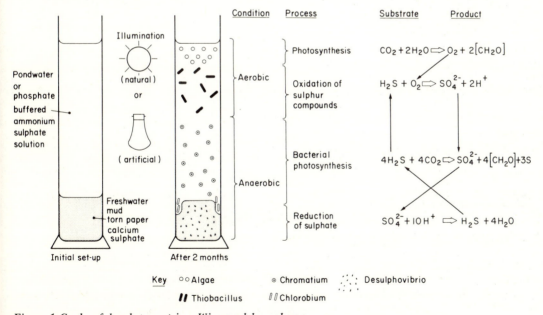

Figure 1 Cycle of development in a Winogradsky column

Ordinary pond water can be used instead of the buffered ammonium chloride solution but the succession does not always then become established.

5 Soil

A healthy soil contains a wealth of organisms among which will be a number of rotting fungi which produce cellulases capable of breaking down cotton threads. One of these, *Chaetomium globosum,* whose black perithecia are often prominent on damp wallpaper, is common in soil from which it may be recovered from strips of cotton buried in the ground for some weeks. If spores of the fungus are placed in flasks with a suitable liquid medium and a range of threads of materials like linen, cotton, terylene etc. are added, students can easily investigate changes in tensile strength with time.

A suitable medium is:

Potassium dihydrogen orthophosphate (KH_2PO_4)	2.5g
Ammonium sulphate $((NH_4)_2SO_4)$	2.0g
Hydrated magnesium sulphate (Epsom salts $MgSO_4.7H_2O$)	2.0g
Potassium chloride (KCl)	0.5g
Calcium chloride $(CaCl_2)$	0.1g
Thiamin	0.002g
Distilled water to 1 litre	

None of these cultural procedures require the use of aseptic technique or the manipulation of agar solidified media. However to obtain mono-cultures both procedures are essential.

STERILISATION

Sterilisation can be effected in a variety of ways.

1 Autoclaves (wet sterilisation)

Whenever possible it pays to use a commercial bench autoclave so that media can be prepared in bulk quickly. A high domed pressure cooker is often adequate for many school purposes but as its volume is small it will inevitably take a lot of time to prepare enough materials for group work by a whole class or classes. Detailed instructions are supplied with cookers but the main points to watch are that all of the air is expelled before adding the weight and that the cooker is not allowed to boil dry. In general the cooker should be allowed to cool naturally rather than by placing it under the tap as this may cause boiling of the medium and subsequent cracking of container bottles.

2 Sterilising ovens (dry sterilisation)

These are used for the sterilisation of glassware such as pipettes, test tubes, flasks and the like. Normally 3 hours at $170°C$ is adequate. This temperature can be achieved in an ordinary domestic cooker.

It is important to remember that some bacterial spores may survive a dry sterilisation process but they are all rapidly killed by the wet process.

Media are not sterilised in an oven because the high temperature causes chemical breakdown of organic components.

3 Flaming

This technique is used to sterilise things like nichrome loops and needles. The articles are held in a Bunsen flame or spirit lamp until they glow dull red over a distance of about 3 cm. Glass spreaders and forceps can be sterilised by dipping in alcohol and flaming off the alcohol. It is important to hold the spreader pointing downwards whilst carrying out this operation or else burning alcohol will run back onto the fingers of the worker.

4 Filtration

This method involves passing liquids through filters which remove bacteria, algae or fungal spores, as the pore sizes are too small for them to pass. It is necessary to use this method with liquids which are heat sensitive and cannot be sterilised by autoclaving. The most frequently used filters are Seitz (asbestos-cellulose) and membrane (cellulose ester). Filters may be sterilised by autoclaving *in situ* in their holders.

5 Ultraviolet radiation

Micro-organisms are killed by radiation of about 260nm (2600 Å) and this is the wavelength used in germicidal lamps. Radiation is used to sterilise the air and working surfaces in inoculating cabinets. It should be noted that these radiations only penetrate to a depth of 1.0 to 2.0 mm in solutions and can only sterilise working surfaces.

6 Steam sterilisation (Tyndallisation)

Materials are poached in a double saucepan, or any two containers which will fit one into the other, for thirty minutes on three separate days. The procedure is used when it is essential to prevent the temperature of the medium rising above $100°C$.

Figure 2 Medical flats and McCartney bottles

7 Chemical sterilisation or disinfection

Apparatus is treated with toxic chemicals, such as hypochlorites, at appropriate concentrations and subsequently washed thoroughly with sterile water.[9]

In addition to sterilisation to prevent extraneous growth simple cleanliness is most important too. Glassware should be free from traces of salts or metals and after normal washing it should be rinsed in glass distilled water before sterilisation.

CULTURE MEDIA

Suitable liquid media can be made by adding water to natural materials like malt (barley grains germinated and then killed), corn meal, vegetable juices or artificial nutrient mixes of natural materials and chemical compounds. They are most conveniently distributed into 'medical flats' or McCartney bottles (see Fig. 2) and sterilised by autoclaving at approximately 103.5kN/m^2 for 15 minutes. Alternatively commercially prepared granules or tablets of extracts can be used.

Solid media are prepared by adding commercially available tablets or granules to glass distilled water. Alternatively raw agar powder can be added to the liquid medium to give a concentration of 1 to 1.5%. The granules, tablets or agar are left to soak for 15 minutes in the water or medium and dissolved, either by steaming for an hour or, more simply, during the autoclaving process. When placed directly into boiling water agar froths up rapidly and displaces the boiling water. This can easily cause severe scalding on the hands of the person carrying out the preparation and so agar should NEVER be added to boiling water nor should boiling water be poured on it. When liquid, the agar medium is distributed into suitable containers, these may then need to be resterilised. Slopes are made by adding about 5 cm^3 of agar medium to a bottle and tilting the bottles before the agar sets. (Fig. 3)

In situations where actual preparation is

Figure 3 McCartney bottles set so as to produce slopes

uneconomic in time, prepared and sterilised media, liquid and solid, are available commercially and can be purchased.

For class practical work a medium is best used as a shallow layer or plate in a sterile glass or plastic Petri dish. The medium to be used is melted and then cooled to 45–50°C, ideally in a waterbath, but alternatively in a saucepan of water with a thermometer to check the temperature. Plates can be poured at higher temperatures but they will take longer to set and will have water of condensation on the lids.

To pour a plate wipe the outside of the warm medium container with a paper towel to absorb water from the water-bath. Remove the cap or plug from the mouth of the vessel and flame it by passing it quickly through a pale blue Bunsen flame. If you are right-handed use your left hand to tilt the lid of the Petri dish and pour about 15 cm³ of the medium quickly in without touching the sides or lid. Close the lid and with the dish on a flat surface, gently rotate it so that the medium agar forms an even layer. Allow the agar to cool and set and then invert the dish to prevent condensation falling onto the solidified medium surface.

When pouring, the lid of the dish should be removed as little as possible to reduce contamination. As far as possible, work near a Bunsen to provide upward air movements—do not work in draughty

conditions or near open windows. If you suspect that the air is contaminated by mould or bacterial spores, then the plates should be poured in a sterile inoculating or transfer cabinet. Alternatively contamination can be reduced by the use of a spray gun (as used to spray garden plants) using a disinfectant solution in water. Spray into the air over the area to be used for pouring plates and swab the area with a clean tissue.

When working with motile bacteria or when inoculating plates with liquid which must dry quickly into the agar, artificial drying is necessary. It can be achieved by a number of methods:

1 storage in the refrigerator for about a week.
2 storage at room temperature for 2 to 3 days (although this risks contamination).
3 inversion of open plates in an incubator. The time taken varies between 10 minutes and 2 hours, depending on the temperature and relative humidity of the incubator.

The plates should be arranged as shown in Figure 4.

Such plates are in fact most conveniently prepared from commercial granules or tablets (e.g. Oxoid), but a home produced form can be made from:

Beef extract (Lemco)	1.5g
Yeast extract	0.1–0.3g
Peptone bact.	1.0g

Sodium chloride (NaC1)	0.5g
Agar	1.5g
Water	100cm^3

It cannot be stressed too strongly that if media are cooled to 45°C, then excess condensation moisture is avoided. Beware—a wet plate can be covered by a motile *Bacillus* overnight at 37°C!

After completing the preparation of a sample of medium it is always worthwhile to carry out a check for sterility. Ambiguous results are thus avoided and the following checks are easily performed:

1 incubate liquid or solid media for 48 hours at 30°C or leave plates at room temperature for 1 week. Any contamination will then be revealed as turbidity or by the presence of discrete colonies. It is desirable to cover wool plugged tubes with aluminium foil so as to prevent evaporation.

2 alternatively include a container of nutrient broth inoculated with bacterial spores, like those of *Bacillus stearothermophilus*, with high heat resistance as a test sample with the batch being treated. If the test sample does not subsequently show any sign of colony development the other items of the batch are almost certainly free of contamination.

MORE SOURCES

With prepared media and agar plates, as opposed to material food substrates, other common natural sources of micro-organisms can be used. Some examples are:

1 Air

No special species inhabit the air. The types which may be found depend entirely upon what particular sources are stirred up in the immediate vicinity of the area being used. In places where there are a number of people moving, or floors are being swept by brooms, dust bearing soil organisms is lifted. Coughs, sneezes and even vigorous talking will distribute nose and throat bacteria. Exposure of nutrient agar plates in a variety of places for 30—90 minutes should, after incubation at room temperature, give an indication of the aerial contamination present. Bacteria, yeasts and fungi are all normally to be found but it is important to appreciate that some of the species obtained may be pathogens.

It is interesting to note that if the air in a room is undisturbed by people or convection currents, all micro-organisms slowly settle and the air becomes virtually sterile. Agar plates exposed in a room first thing in the morning will thus show very different growths to those exposed after a room has been used by groups of students.

2 Water

Natural waters are likely to contain faecal material from a variety of organisms and a range of organic substrates which are ideal foods for micro-organisms. Colonies may be obtained either by direct plating of natural water samples or alternatively, if there are few bacteria present, then the water can be filtered through sterile membranes which are then incubated on nutrient agar plates.

3 Soil

The number of micro-organisms in soil samples varies very considerably. A rich garden loam can contain as many as 5.0×10^8 bacteria per gram, whilst a soil containing large amounts of clay or sand may be relatively barren. Sub-soils have poor micro-organism populations too so that if a piece of wood is

Figure 4 Plates arranged on an oven shelf for drying

11

buried in the top 25 cm of a soil it rots quickly, but over a metre down it will remain unattacked for long periods of time.

Soil crumbs can be placed directly on nutrient agar (see page 10) or malt agar (see Chapter 3, page 42). Alternatively, soil can be diluted with water and the dilutions spread on plates. Isolation of sporing bacteria can be made by boiling soil for a few minutes to destroy non-sporing forms and then subsequently spreading it onto plates for spores which have survived to develop. Once a mixed source of organisms has been achieved it becomes useful to isolate mono-cultures by inoculating selected species into broths or onto plates.

INOCULATING

For inoculating tubes or flasks plugged with wool, an inoculating loop is normally used. This consists of nichrome wire inserted into a special metal holder (Fig. 5). Twenty to twenty-three 'standard wire gauge' is ideal and the finished loop should be about 5.0 cm long. The loop is sterilised and allowed to cool. The cotton wool plug is removed from the tube or flask by grasping it between the little finger and the palm of the hand. Where appropriate, the neck of the culture container is flamed and the loop intro-duced into the container. The material is picked up, removed from the container and the top replaced. The organisms are then transferred to the new culture vessel by dipping the loop into the fresh broth.

Bacteria picked up on a loop can also be 'streaked out' on the surface of agar medium in Petri dishes such that at the end of the operation single cells are well separated and produce single colonies from which a pure culture can be derived. The morphology of such colonies is important in the identification of bacteria. 'Streaking out' on plates is normally an essential process in obtaining a pure culture of an organism from a mixed population. The usual method of streaking is illustrated in Figure 6. The concentrated culture is streaked along one side of the plate (I). The loop is next sterilised and organisms from zone (I) are streaked out on line (II). The process is repeated twice more along (III) and (IV). In zone (IV), single cells should be left, which will be fairly well scattered.

The loop can be cooled between operations by plunging it into the agar at the edge of the plate. Note that Petri dish cultures should be incubated lid downwards so that droplets of condensation collect in the lid and not on the agar surface.

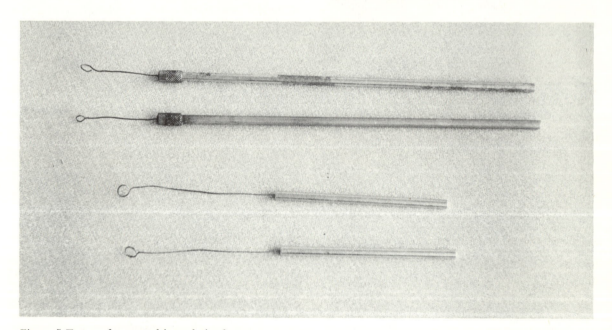

Figure 5 Types of mounted inoculating loops

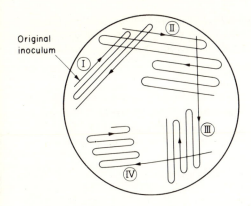

Original inoculum

Figure 6 A sequence for streaking out on a Petri dish

SAFETY PRECAUTIONS

In order to reduce contamination and to avoid health hazards, it is advisable to observe the following general safety precautions[9,11,13,22,23]:

1 all organisms isolated should be regarded as potential pathogens and handled with caution.

2 avoid isolating microbes from potentially dangerous sources such as human mucus, pus from cuts and faecal material. Other natural sources such as polluted water can contain pathogens.

3 cultures should always be kept in sealed containers. Petri dishes should be sealed together with self-adhesive tape across the lid and base. If inspection without a lid is essential the culture should be killed by placing a few drops of 40% formaldehyde solution on filter paper inside the cover, at least one hour but preferably twenty-four, before examination.

4 avoid as far as possible the production of aerosols which result, for example, from flaming a loop of bacterial suspension in a bunsen. Always immerse the loop in boiling water before flaming.

5 laboratory coats should be worn at all times.

6 pencils and pen handles should never be placed in the mouth, nor should labels be moistened with the tongue.

7 pipettes should be filled using rubber or plastic teats or other pipette fillers and never placed in the mouth.

8 all cuts should be protected with adequate waterproof dressings before starting work.

9 hands must be thoroughly washed before leaving the laboratory.

10 nobody should eat, drink or smoke in the laboratory.

MANIPULATIVE PROCEDURES

Since pure cultures of single species of microbes are normally chosen for study, it is important to prevent air-borne organisms being introduced into the material being studied[13,23]. The following points of manipulative procedure will restrict contamination:

1 the inner surfaces of cotton wool plugs and other stoppers must never touch any object apart from the inside of the culture container.

2 before using the culture, withdraw the plug or stopper and flame the mouth of the container in a Bunsen flame.

3 any instrument introduced into the culture container must be flame sterilised first. Thus wire inoculating loops are heated to redness and the end of the handle passed through the flame.

4 keep culture tubes and plates as horizontal as possible when open—this minimises the risk of aerial contamination.

5 try to work in draught-free conditions and avoid making air disturbances when handling microbes.

6 it is a good idea to work close to the Bunsen so that up-draughts prevent organisms falling into culture containers.

7 it is best to use a small inoculating or transfer cabinet in laboratories, particularly where contamination is bad. (See Fig. 21 page 43.) The air inside the cabinet is sterilised by an ultraviolet lamp or else it is fed with filtered air under pressure. The inner working surfaces are best cleaned and sterilised with a disinfectant solution like 5% 'Domestos', 70% ethanol or 1—2% surface active disinfectant, for example Harris 'BAS Cleaner'.

8 a culture should never be opened for longer than is necessary to take samples. Plugs, stoppers or lids must be replaced as soon as possible.

9 cultures spilled on bench, floor or person must be swabbed immediately with a disinfectant solution.

10 all instruments must be sterilised after manipulating micro-organisms to prevent their dissemination throughout the laboratory. Contaminated glassware should be placed in a disinfectant solution like 5% 'Domestos' or hypochlorite before being washed.

11 cultures to be disposed of should be sterilised in an autoclave or a pressure-cooker before the glassware is washed out. Screw lids must be partly unscrewed before autoclaving. The agar in bottles, tubes or dishes should be poured out into plastic bags just before it sets and the containers washed out immediately with hot water. The plastic bags must be sealed as soon as possible and may then be disposed of in refuse disposal bins. After sterilisation, broths may be poured into a sink and flushed away with water.

12 plastic Petri dishes should be put into autoclavable disposal bags which are then sealed and autoclaved. Afterwards the unopened bags may be placed in refuse disposal bins. Alternatively the dishes may be placed in stout waterproof bags and incinerated in a furnace. Since they are made of polystyrene they burn safely, producing carbon.

KEEPING STOCK CULTURES

Having collected, or purchased, a range of micro-organisms, it is worthwhile trying to preserve cultures. There is no one procedure which guarantees success in all places; rather the individual worker has to try a range of possibilities to find those which best fit his own situation.

Fungi and bacteria

Some possible methods for fungi and bacteria are:

1 in screw top bottles—slopes

A number of bacteria remain viable on slope cultures if the tops are screwed down to prevent dessication. Fungi can also be kept by this method but some die quickly unless the screw top is left loose. Certain fungi can be kept by allowing the slopes to dry naturally in a dark place at room temperature. Otherwise cultures in general should be kept at $4°C$.

2 in screw top bottles—stab cultures

An excellent method of preserving bacterial cultures is to stab them into soft nutrient agar (identical to nutrient except agar is at 0.6%. See Chapter 1, page 11). Small tubes with plastic stoppers can be used but have the disadvantage that the top cannot be autoclaved and are sterilised by dipping in alcohol and allowing the alcohol to evaporate off. Alternatively 7 cm^3 (¼ oz) bottles with screw top can be used in a similar manner but the top

must give a water tight seal. Cultures kept by this method can be viable for 2—3 years. The method has been used particularly for strains of *Escherichia* and *Salmonella*.

3 broth culture

Bacteria and yeasts can be kept for several months in broth cultures kept at $4°C$. Cultures kept in this way are subject to contamination and should be purified by single colony isolation before use.

4 deep freeze

Slopes or liquid suspensions of cells can be placed directly into the deep freeze. The lower the temperature the quicker the freezing will take place. Cells may be frozen as a pellet or as a thick suspension in distilled water or a nutrient medium.

5 freeze dried cultures

In general, this requires complicated equipment and is not suitable for school conditions. Frequently, however, cultures received from suppliers are freeze dried. When opening them it is important that the directions which come with the ampoule should be carefully followed. A typical example of such directions is:

1 make a file cut at the midpoint of the cotton wool in the ampoule and crack the tube to allow air into the ampoule *slowly*.

2 remove the numbered paper strip with flamed forcep to a nutrient slope.

3 add nutrient medium to the tube and streak a loopful on a plate.

4 incubate the ampoule with a fresh sterile cotton wool plug and add more medium to provide further inoculum.

Unused, unopened ampoules should be stored in a dark cool place.

6 silica gel dried cultures

Cellular metabolism in a micro-organism can be almost completely stopped by adding a suspension of the organism in skim-milk directly to anhydrous silica gel so that long term storage is possible.[38],[46] This method has been used to preserve fungal cultures for some years and it is also successful with a number of bacteria like *Escherichia* and *Micrococcus*. Unfortunately viable cells of algae like *Chlamydomonas* and *Euglena* cannot be recovered from dehydrated cultures.

Equipment

Suitable screw-cap containers to allow repeated

14

sampling of the gel. The cap liners should be of metallic foil, as waxed paper or cork liners discolour the gel on heat sterilisation. Non-fat skim milk powder like Marvel or other proprietary brands. Purified silica gel without indicator (6—22 mesh).

Setting up

The following procedure is based on the use of screw cap vials containing 4g of gel.

1 Threequarters fill containers with gel and plug with cotton wool. Sterilise in dry heat for a minimum of 90 minutes at $180°C$. Tubes must be stored in a dry atmosphere and if they become damp it is necessary to resterilise. If the tubes have bakelite tops sterilise these separately by autoclaving at approximately 69 kN/m^2 for 10 minutes.

2 Prepare a 5% solution of milk in distilled water for fungi and a 15% solution for bacteria. Distribute 2.0 cm^3 lots to small tubes. Autoclave at approximately 103.5 kN/m^2 (or approx $121°C$) for 10 minutes and store in a refrigerator at $4°C$.

3 Grow fresh well conidiated slopes of strains of fungi or fresh, vigorous bacterial colonies in broth.

4 Stand the silica gel containers in an ice bath for at least 30 minutes.

5 Tip milk on to slope and make a heavy suspension of fungal conidia using a long sterile wire to scrape off the conidia. (Purists may prefer to use pipettes to transfer the milk and the conidial suspensions, but the pipettes are difficult to clean.) Centrifuge bacteria from broth and resuspend them in the milk solution.

6 Tip micro-organism suspension $(10-1.5cm^3)$ onto cold gel, return container to ice-bath at once and keep it there for at least 15 minutes. (Considerable heat is evolved when the gel is wetted, hence the necessity to cool the gel and use cold milk.) Do not saturate the gel, it should only be about threequarters wetted.

7 Keep gels at room temperature until the crystals readily separate when shaken (about a week).

8 Check a sample for viability.

9 Screw caps down firmly. Store over indicator gel in an airtight container at $4°C$. Plastic 'freezer' boxes with a good seal are ideal. The indicator gel will require drying once or twice a year.

10 Sub-cultures are taken by transferring a few crystals of gel to a suitable agar slope.

Note The gels can be kept at room temperatures for prolonged periods without any apparent loss of viability. It is convenient to mail cultures by sending a few crystals of gel in a small vial.

Algae and most protozoa; biphasic cultures

Algae like *Euglena, Chlamydomonas, Pandorina, Zygnema* and *Vaucheria,* and ciliates like *Paramecium,* and *Spirostomum* are most readily kept in continuous culture in a biphasic system of soil and water.[55,61,62] Almost any glass vessel can be used but the most useful are boiling tubes, jam jars and tall 250 cm^3 beakers. A little good garden soil is first put into the vessel to a height of not more than 1.0 cm.

Next water is added gently to within 2.5 cm of the top of the vessel so that the final mix is not cloudy. The volume of water added should be at least seven to ten times the volume of the soil and if the final medium is anything darker than a light golden brown less soil must be used. The culture vessels are plugged with cotton wool or covered with greaseproof paper or half of a Petri dish and then sterilised in a steamer for one hour on each of two consecutive days. Very large vessels will need to be steamed longer.

In most places tap water can be used because chlorine or any other volatile contaminant will be driven off when the medium is steamed and any harmful traces of metal like copper or lead will be absorbed by the soil. If distilled water is used it is important to bear in mind that water from a badly managed metal still can be more harmful to micro-organisms than tap water!

Finally, the culture vessels are allowed to stand for 24 hours and then inoculated when oxygen and carbon dioxide have re-entered the water from the air.

It is best to collect a large soil sample to start with, say 0.25 m^3, and pass it through a garden sieve to remove large stones and roots. It should then be air dried and tested to see if it is satisfactory. A suitable sample may be kept for many years but if you collect a second sample from an identical spot you cannot rely on getting the same results!

Algae need light, but direct sunlight is to be avoided. A north facing window is ideal or 80W warm white fluorescent tubes stimulate good growth in cultures 30—60 cm away.

Ciliates need a bacterial flora to feed upon and this is encouraged by the addition of pearl barley, wheat or rice grains (2 per boiling tube; 4—6 per jam jar) beneath the soil before steaming.

The soil in a biphasic cultural medium of this type acts as a reservoir slowly releasing substances as they are required and also absorbing any harmful waste products that may accumulate. It also has remarkable buffering and chelating properties which adjust to any sudden chemical disturbances.

Free-living amoebae

Free-living amoebae need to be treated as a special group on their own. All methods are empirical and depend upon experience and the maintenance of several cultures for success. They are conveniently grown in large Petri dishes kept in the dark at 20°C containing either Chalkley's or Pringsheim's medium with 2 or 3 previously boiled wheat or rice grains.[59,64]

Chalkley's medium

Calcium chloride ($CaCl_2$)	0.06g
Sodium chloride (NaCl)	1.0g
Potassium chloride (KCl)	0.04g

Glass distilled water 1 litre
Dilute 10 times with water before use

Pringsheim's medium

Solution A:

Calcium nitrate ($Ca(NO_3)_2.4H_2O$)	2.0g
Iron (II) sulphate ($FeSO_4.7H_2O$)	0.02g
Magnesium sulphate ($MgSO_4.7H_2O$)	0.2g

Glass distilled water 500 cm^3

Solution B:

Dipotassium hydrogen phosphate (K_2HPO_4)	0.2g
Sodium chloride (NaCl)	0.2g

Glass distilled water 500 cm^3
To use, mix 5.0 cm^3 of Solution A with 50.0 cm^3 of Solution B and make up to 1 litre with water.

In either case use glass distilled water as water distilled from copper is poisonous to all but a few resistant organisms. The solutions should be stored in brown bottles. Shake the bottles thoroughly before diluting. Measuring-cylinder accuracy is enough.

To inoculate, shake the original culture bottle well to remove amoebae from its walls. Throw all of its contents into one dish with a lid and leave the dish overnight to settle. Put the dish on to a sheet of black paper, and using a stereomicroscope (x10) and strong lateral illumination, examine the growth. The animals should be so numerous that many of them can be seen wherever one looks.

This being the case, put the lid on, and holding the dish flat on the bench, shake it vigorously from side to side. Now pour off one half of the original into a second dish. Add culture medium to each dish to a total depth of not more than 1.0 cm. Divide the old wheat grains between the two. Add new grains to a total of 4 or at most 5. Set both cultures aside and leave them *undisturbed* for four weeks. During this time the new wheat grains will grow beards of the fungal mycelium that should have been visible on the original grains. The amoebae will climb into the mycelium, where the supply of bacteria and of the micro-organisms that feed on them is greatest. Left *undisturbed* the amoebae will eventually form a continuous snowy mass on the grain, with numerous out-liers in the mycelium around and on the floor beneath.

Once satisfied that the growth is enough (this can only be by rule of thumb: one's judgement improves with experience) subculture, repeating the process as before. Alternatively, using flamed forceps, lift out a well-populated wheat grain without squashing it, and transfer it to another prepared dish, adding more wheat grains as before.

In some cultures the wheat-grains do not grow mycelial beards. Why this happens is not known. Such a culture will not give good supplies of amoebae: it is best to discard it. When a culture develops an irridescent sheen on its surface, it is losing health. Subculture it as soon as possible, or add more medium, removing beardless and empty (brown, translucent) grains and adding more. Should the surface develop a white scum (this grows from the irridescence) cut a sheet of clean blotting paper to fit across the dish and wipe the surface away, drawing it up the side of the dish; then pour off one half of the old medium and replace it with fresh and renew the wheat grains.

Chapter 2 Working with bacteria

A moderate sized rod-shaped bacterium is about 3 μm long and nearly 1 μm wide. The majority of students find it difficult to appreciate exactly how small this is and a true appreciation of dimensions of this order only comes as a result of experience in handling the cells. Fortunately bacteria make up for their small size by their vast numbers and so they are not difficult to find.

The part played by bacteria in causing disease is well known but many people overlook the action of bacteria, aided by the fungi, in causing decay of organic material. The putrefaction process both removes bodies and returns minerals, especially nitrates, to the earth where they are taken up by plant life. The bacteria are thus destroyers and universal providers for other living things. Since most are colourless and small, bacteria are extremely difficult to see in simple wet preparations and so it is necessary to prepare stained films or smears.

SOME METHODS FOR STAINING BACTERIA

Preparation of bacterial smears

It is essential that microscope slides used for making smears should be perfectly clean and free from grease; otherwise uneven films will result. To remove all traces of dirt the slides should be soaked in chromic acid for three days, rinsed thoroughly in distilled water and then stored for subsequent use in wide necked jars containing industrial methylated spirit.

A suitable chromic acid mixture can be prepared from:

Potassium dichromate	63.0g
Concentrated sulphuric acid	960.0cm^3
Distilled water	35.0cm^3

The dichromate is dissolved in the water in a 2 litre beaker standing in a trough of cold water. The concentrated sulphuric acid is then added VERY CAREFULLY, a little at a time, to the beaker. The resulting chromic acid mixture is HIGHLY CORROSIVE and must be used with extreme care. It will work until it becomes green in colour.

In order to use stored slides they are removed with forceps, excess alcohol is allowed to drain back into the storage jar and they are flamed to remove the remaining alcohol.

The essential steps in actually making a smear are:

1 to make a smear from a colony on solid medium, place a small drop of sterile water on a slide with a loop, add a little of the bacterial growth to the drop and emulsify by gently stirring with the loop. The drop should appear faintly turbid. Spread the suspension evenly over the slide. If the suspension does not spread but collects into droplets, the slide is greasy and should be discarded. In the case of broth cultures, one loopful is taken up with a cool sterile inoculating loop and spread over a fairly small area on the slide.

2 allow the slide to dry in the air or hold it high over a non-luminous Bunsen flame.

3 mark the surface of the slide bearing the smear with wax pencil or a glass knife.

4 fix the film of bacteria by passing the dried slide two or three times through the flame with the filmed surface upwards. The slide should be just too hot to be borne on the back of the hand. Fixation causes the bacteria to adhere to the slide and makes the organisms stain more readily.

Staining methods[4],[7]

1 Methylene blue, a useful general stain

1 Place the heat fixed smear on a staining rack, made out of two pieces of glass rod joined at the ends with rubber tubing (see Fig. 7), over a sink and flood it with Loeffler's methylene blue and leave for at least 5 minutes.

To prepare Loeffler's methylene blue take:

Methylene blue	0.5g
1% Potassium hydroxide solution	1 cm^3
Ethanol	30 cm^3
Distilled water	100 cm^3

Warm the water to 50°C, stir in the methylene blue and add other ingredients. Filter before using.

2 Wash the methylene blue off with water. Although fixed films adhere closely to slides, direct jets of tap water may dislodge them. Washing should be

17

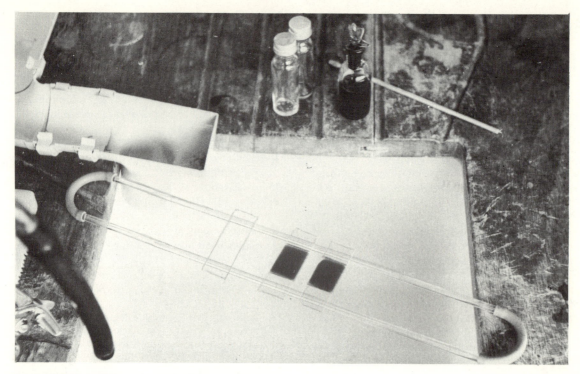

Figure 7 Slide staining rack made from glass tube

done in beakers or in gently flowing water.

3 Dry the slide by blotting between two sheets of blotting or filter paper.

4 Place a small drop of immersion oil over the smear and examine it using an oil immersion lens.

2 Gram staining bacteria

This is a differential double-staining method which forms the basis of most examinations and the preliminary identification of bacteria. The bacteria are first stained with methyl violet and then treated with iodine which 'fixes' the stain. The smears are next treated with alcohol, which entirely removes the violet stain in the case of gram-negative bacteria. These decolourised bacteria are counterstained with dilute carbol fuchsin or any other contrasting counterstain. This method divides bacteria into two classes:

gram-positive—those which appear purple since they are not decolourised by the alcohol, e.g. *Bacillus*.

gram-negative—those which appear pink since the alcohol removes the methyl violet and they are only

18

stained by the carbol fuchsin or whatever is the colour of the counterstain used e.g. *Escherichia*.

Some organisms are gram-variable and some gram-positive organisms may lose the gram staining property in old cultures.

Lugol's iodine for fixing is made from:

Iodine	1g
Potassium iodide	2g
Distilled water	100cm^3

Add the iodine and iodide to about 25 cm^3 of the water and when all the material has dissolved add in the remainder of the water.

Methyl violet is made from:

Methyl violet	2g
95% Ethanol	20cm^3
1% Ammonium oxalate solution	80cm^3

Dissolve the methyl violet in the ethanol and add the ammonium oxalate solution which has been made up in distilled water. The mixture should stand for two days before use.

Dilute carbol fuchsin is made by diluting 10 cm^3 of

saturated alcoholic fuchsin with 40 cm^3 of distilled water.

A convenient procedure for gram staining is:

1 prepare a heat-fixed smear from an 18—24 hour old culture.
2 stain with methyl violet (1—2 minutes).
3 wash with tap water and drain off excess tap water.
4 cover the slide with Lugol's iodine for 1 minute.
5 holding the slide on a slant, drip alcohol over it until no more blue colouration is seen to leave the smear (only 5—15 seconds in the case of well-prepared thin smears).
6 wash immediately with tapwater and drain.
7 counterstain with dilute carbol fuchsin or other contrasting stain (1—2 minutes).
8 wash the slide well, blot dry and examine under oil immersion.

When preparing slides, the stains used often run off the ends onto the fingers of the worker. This can be prevented by drawing two lines about 1 cm from each end with wax pencil as in Figure 8.

Wet mounts—hanging drop preparations [A.L.]

Although skill is needed to set them up, these are useful for examining motile bacteria and other micro-organisms. A loopful of the organisms in suspension is placed on a coverslip, which is then inverted over a cavity slide, so that the drop is suspended in the cavity. Spots of vaseline are placed on each corner of the cover glass to hold this on the slide. Organisms can be observed under the high power objective; the oil immersion lens is not usually required.

It should be remembered that non-motile organisms of the size of bacteria show Brownian motion. Motile organisms, in contrast, move in a definite direction relative to each other. It is often useful to reduce the amount of light coming through the microscope to improve the visual contrast of organisms in hanging drops.

DESTROYING BACTERIA

Physical antimicrobial treatments

1 Autoclave treatments

Steam under pressure is much more effective than dry heat at the same temperature in killing bacteria. This is due to the difference in the effect of heat on wet or dry protein in the cells and so moist heat sterilisation (autoclaving) can be applied at temperature and time combinations much lower than when dry heat (oven sterilisation) is used. Articles for sterilisation need to be in a hot-air oven at 160°C for 1.5 hours but by using an autoclave to steam under pressure, forces of approximately 69kN/m^2 or 103.5kN/m^2 are effective in 15 minutes.

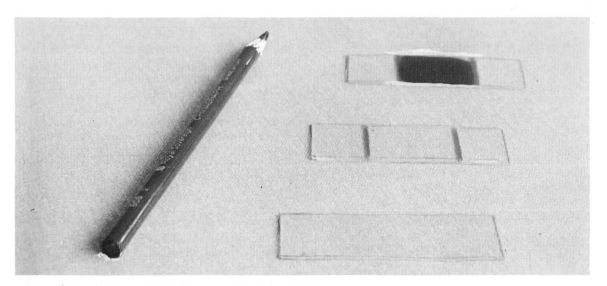

Figure 8 Slide with wax pencil lines to avoid stain spreading

To demonstrate some of the differences which water can make to effectively killing bacterial spores, garden soil (air dried at 37°C) and *Bacillus globigii* spore suspensions (about 10^8 spores per cm³) can be used. Take filter paper strips, double concentration nutrient broth (using twice the quantities given in Chapter 1, page 10, but omitting the agar), four dry medical flat bottles (227cm³ (8oz) are convenient) and some sterile 0.85% sodium chloride solution (saline). Then:

1 to each of the strips of filter paper (placed on a glass slide) add one drop of the *Bacillus globigii* spore suspension.
2 after drying in air, insert a strip into each of two flat bottles (label one A and the other B).
3 place in each of the other two flat bottles 1—2g of dried garden soil (label one C and the other D).
4 pipette 10cm³ of saline into bottles A and C.
5 with the screw caps half a turn from the fully closed position, autoclave bottles A and C at approximately 103.5kN/m² for 10 minutes.
6 allow the bottles to cool below 45°C and then pipette 10cm³ of saline, *aseptically*, into bottles B and D.
7 pipette 10 cm³ of sterile double-strength nutrient broth into all bottles.
8 screw down bottle caps loosely and incubate at 30°C for 4—5 days. Then inspect for growth by checking for turbidity.

Bacillus stearothermophilus is a spore producing bacterium which will not grow at temperatures below 55°C (the best culture temperature is 56—60°C). Its gram-positive, motile rods produce oval spores which are heat resistant and will survive autoclaving at approximately 69kN/m² for 10 minutes.

As the spores of this organism are especially resistant to heat, absorbent paper strips impregnated with a spore suspension are used to test the effectiveness of a sterilisation procedure. Dry treated strips are placed inside the apparatus and, after sterilisation, the strips are removed and incubated in nutrient broth for 4 days at 55°C to see if any turbidity develops as a result of any spores surviving the oven or autoclave conditions.

2 Ultraviolet light

The effect of ultraviolet light on bacterial cells is readily illustrated by using a nutrient broth culture of *Chromobacterium lividum*. The gram-negative motile rods of this organism produce a water

20

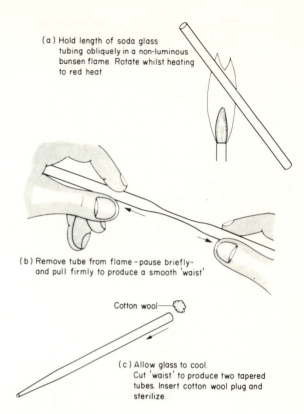

(a) Hold length of soda glass tubing obliquely in a non-luminous bunsen flame. Rotate whilst heating to red heat

(b) Remove tube from flame – pause briefly – and pull firmly to produce a smooth 'waist'

Cotton wool

(c) Allow glass to cool. Cut 'waist' to produce two tapered tubes. Insert cotton wool plug and sterilize.

Figure 9 Pulling a pasteur pipette

insoluble violet pigment which makes them easy to see and the production of pigment is better between 22°C and room temperature than at an incubator temperature of 37°C.

Take five nutrient agar plates, a glass spreader in alcohol, three sterile thin sheets of metal which will cover a Petri dish and a pasteur pipette fitted with a rubber teat.

Pasteur pipettes are easily made by taking 15.0 cm lengths of about 6.0 mm diameter soda glass tubing, plugging the ends with non-absorbent cotton wool, and then 'drawing out' the centre in a Bunsen flame. To do this hold the middle of the tube in the flame until it becomes red hot, rotating it all the time. Remove it from the flame and pull it out to about twice the original length (Fig. 9) and, when cool, cut the centre of the neck produced. Finally, sterilize the pipettes in sets in an autoclave.

Having assembled the necessary equipment:

1 by means of the Pasteur pipette, deposit 5 drops of *Chromobacterium lividum* culture on the surface of each plate.
2 ignite the alcohol on the glass spreader and after allowing a few seconds for it to cool, spread the culture evenly over the surface of the plates (you need not flame the spreader between plates.)
3 replace the lids of three plates with the metal sheets, and the lid of a fourth plate with a glass lid if plastic dishes are used.
4 expose all four plates to ultraviolet light at about 15—20 cm from the source as follows by removing the metal plates one at a time:

Plates	1 (control—	2	3	4	5 (glass lid
Time in	not treated)				removed)
minutes:	0		0.5	1	3 3

5 refit the lids and incubate the plates at 30°C for three days.

It will be found that the growth on exposed plates is inversely related to the time of exposure. The plate with the glass lid will show no effect since glass absorbs ultraviolet radiation. This experiment can also be carried out in sunlight if plates are exposed in the sun for over an hour, depending on the season of the year.

Chemical antimicrobial agents

1 Disinfectants

There are a wide range of poisonous substances which will kill bacteria and the popular press normally carries dramatic advertisements extolling the virtues of brand-named disinfectants. Some of the claims made with respect to these bacteriocides may be over-stated for, unfortunately, no one compound is ideal for all bacteria in all circumstances. The pale green gaseous element chlorine is a most effective bacteriocide on its own. It works well in swimming pools or in drinking water but it is virtually useless on contaminated surfaces, like dirty floors, since it all too readily combines with other chemicals present and so loses its disinfectant properties.

Although caustic, carbolic acid or phenol has been used as a disinfectant for many years. The effect of treatment with 5% phenol solution, for different periods of time, on the common bacterium *Escherichia coli* can be readily shown.

Take two dried nutrient agar plates, 9.0 cm^3 of distilled water in each of two small conical flasks, a nutrient broth culture of *E. coli* and a 5% phenol solution labelled 'caustic, handle with care'. Then:

1 divide the under-surface of the base of both of the plates into four sectors with a wax or chinagraph pencil and mark one 1 minute, one 5 minutes, one 10 minutes and one 30 minutes. Mark one plate control and the other disinfectant.
2 transfer two loopfuls for the *E. coli* culture into each conical flask and shake.
3 pipette into one conical flask, marked disinfectant, 1 cm^3 of 5% phenol solution. Shake.
4 take a loopful from each conical flask and streak it out in the respective sectors of the disinfectant or control plates at times of 1 minute, 5 minutes, 10 minutes and 30 minutes.
5 incubate the plates at 30°C for 24—48 hours and observe the amount of growth.

There is some variation in the effect of phenol solutions on different strains of *E. coli* and experience of a particular culture may show that it is necessary to modify the exposure times given here in order to improve the response differentiation against time.

2 Antibiotics

The antibacterial compounds (antibiotics) produced by living organisms, like the fungus *Penicillium notatum*[40], differ from disinfectants in that they exhibit a selective action at certain sites in the cell (e.g. ribosomes, nucleus or plasma membrane). Penicillin is thought to act as an antagonist to N-acetylmuramic acid, a component of the bacterial cell wall. Thus its action depends on the cell wall composition and it affects mainly gram-positive bacteria (it can be said to have a narrow spectrum). Tetracyclines act by chelating metal ions essential for the activity of many enzymes. They are active against a wide variety of bacteria (they have a broad spectrum effect). Nystatin is an antibiotic whose action depends on its combination with sterols in the cytoplasmic membrane. It is active against fungi but bacteria which do not have sterols in their membranes are not sensitive to its action.

The effects of these antibiotics can be investigated by taking nutrient broth cultures of a *Staphylococcus* sp. for example *S. epidermidis*, *E. coli* and bakers' yeast *Saccharomyces cerevisiae*, three dried nutrient agar plates, antibiotic treated paper discs (penicillin, tetracycline, nystatin) which are available commercially, Pasteur pipettes and glass spreaders.

Then:
1 add 5 drops of each separate culture to one of the nutrient agar plates using a Pasteur pipette and spread them out with a flamed spreader.
2 with flamed forceps, place one each of the three antibiotic discs about 1.0 cm from the edge of the agar medium.
3 incubate the plates at 30°C for two days and then make tracings of the plates onto millimetre graph paper so as to record the diameters of the zones of inhibition.

It is interesting to repeat the investigation with brand disinfectants using them neat, at the stated concentration, and much diluted as they are frequently used in the home.

PERSONAL HYGIENE

The human body has a number of defence systems against serious pathogenic bacterial invasion. This is just as well since in addition to carrying a considerable natural population of micro-organisms internally in our gut, our skin also houses a potentially more dangerous bacterial flora.

A demonstration of the existence of bacteria on our hands can almost be described as an educational classic. Although the tough skin of the hand protects its component cells these bacteria are easily transferred to more sensitive membranes around the eyes or to the mouth and onto food. Washing is important.

Finger prints

When culturing micro-organisms from human body surfaces pathogens may be found and STRINGENT PRECAUTIONS must be taken to avoid infection—see Chapter 1, page 13.

The presence of bacteria on fingers is most readily demonstrated by preparing a nutrient agar with a lactose/neutral red indicator.[16] 'Fermenters' in the bacterial population convert the lactose to lactic acid and their positions are clearly demarcated by a colour change where the pH has been lowered.

To prepare the indicator solution take:

Lactose	10 g
Neutral red	1 g
Distilled water	100 cm^3

Dissolve the solids in the water and autoclave the mixture at approximately 103.5kN/m^2 for 15 minutes. Two cm^3 of indicator solution are added

22

with a sterile pipette to 100 cm^3 of melted and cooled nutrient agar prior to pouring the plates used for finger impressions. The medium is mixed thoroughly but gently to minimise the formation of bubbles and plates are poured in the normal way and dried.

The finger impressions are made by gently rubbing the pads of the fingers and thumb on the surface of the agar of a plate with a rotational movement (see Fig. 10). After preparing plates from unwashed and washed hands, they should be sealed with Sellotape, incubated at 37°C for two days and then examined.

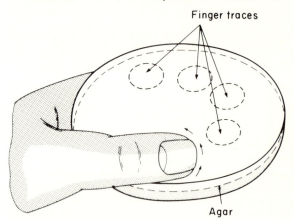

Figure 10 Rotating fingers on an agar plate

Toilet training

An impressive demonstration of the importance of washing hands after using the toilet and the real meaning of notices of the kind 'Have you washed your hands?' can be done with the pigmented bacillus *Chromobacterium lividum*.

Take a nutrient agar plate, divide the under surface of the base into three sectors with chinagraph pencil and label them 1, 2 and 3 (Fig. 11). Dip a swab of cotton wool in a *C. lividum* broth to saturate it and place it in a sterile Petri dish. Then:
1 to protect the skin put on sterile rubber gloves with rough finger tips and gently rub the surface of the forehead with a circular motion to give the glove tips a normal body bacterial flora.
2 touch the cotton wool swab with the right gloved thumb and press your thumb onto sector 1 of the agar plate.
3 cover your left gloved thumb with a piece of toilet tissue. Press this thumb onto the swab, discard the

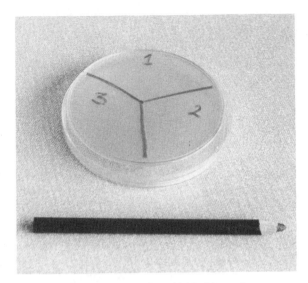

Figure 11 Nutrient agar plate divided into three segments by chinagraph on lid

toilet paper and then press this thumb on sector 2.
4 wash your gloved hands thoroughly with soap and water and press one of your thumbs on sector 3.
5 remove the gloves carefully so as to avoid contaminating the fingers and place them in an appropriate container for resterilising.
6 incubate the 'thumb' plates at $30°C$ overnight and examine. (Of course throughout the investigation great care must be taken to avoid touching the face or body with the gloved hands.)

Students are often interested to carry out a 'Which' type survey of different brands of toilet papers and to compare the effect of using two or more thicknesses of paper on the left thumb.

FOOD BACTERIOLOGY

In order to maintain food supplies for man, the food technologist is frequently at war with bacteria. The food substrate is equally nutritious to the bacterial cell as well as to man and so it is necessary to take steps to preserve foods and prevent microbial spoilage. However, some fermentations induced by micro-organisms produce new foods and at the same time generate a preservative compound (e.g. in yoghurt production lactic acid is formed and this inhibits the growth of other organisms).

Few systems can be guaranteed to be perfect and the food industry frequently uses the presence or absence of bacteria as a means of 'quality control'.

Preservation of foods

Food substances can be preserved by the addition of materials which will inhibit the growth of micro-organisms (preservatives) or they can be rendered sterile by heat or radiation.

Depending upon the intensity of heat treatment (the time/temperature combination) food can be partially freed or become virtually free of micro-organisms. The severity of treatment is determined by considerations such as the nature of any pathogens likely to be present and their heat resistance; nutritive deterioration of the food; the acceptability of heated food; the anticipated extension of storage life; and, not least in a commercial industry, cost.

1 Heat applied to milk (pasteurisation)

In the pasteurisation of milk the aim is two fold: firstly to free milk from actively growing spoilage bacteria (e.g. lactic acid bacteria and coliforms) so that its life is extended for a few days under refrigeration, and secondly, to free it from pathogens (e.g. any tubercle bacilli).

A raw milk sample, kept at room temperature overnight, can be pasteurised by:
1 heating 100 cm^3 raw milk in a 500 cm^3 conical flask for 30 minutes allowing an extra 5 minutes for heating up) at $63°C$ in a water bath. The flasks should be clamped and immersed so that the water level is above the milk level.
2 cool in tap water and store.

Milk pasteurised in this way is said to have been treated by the 'Holder method' or the 'low temperature, long time' (LTLT) procedure. An alternative is the 'high temperature, short time' (HTST) process in which the raw milk is heated to $72°C$ for 15 seconds. Both methods kill all pathogens in milk but the quicker HTST process has now generally replaced the 'Holder method' in large commercial dairies.[44]

2 Use of inhibitory materials

High concentrations of sugar or salt with food inhibit the growth of micro-organisms by causing strong osmotic effects. In the case of sweetened condensed milk, the high concentration of lactose and added sucrose preserves it until it is diluted.

23

This is easily demonstrated by taking:
a can of sweetened condensed milk,
a wide-bore 10 cm^3 pipette, and
a nutrient agar slope culture of *Streptococcus lactis* (or commercial fermented buttermilk).
Then:

1 keep the can in a water bath at 45°C for 15 minutes.
2 dry the can with a cloth, then ignite alcohol on one end and open it with a flamed punch or can-opener.
3 transfer 5 to 10 cm^3 into a sterile tube.
4 transfer 2 cm^3 into another tube containing 9 cm^3 sterile water.
5 inoculate both tubes with the culture of *Streptococcus lactis* and incubate at 30°C for 2—5 days.
6 finally, determine the relative amount of growth in each tube by means of smear preparations stained with methylene blue (growth will be noticed only in the diluted sample).

QUALITY CONTROL

The presence or absence of bacteria in a food product is estimated in different ways for different foods.

With milk, counts of viable bacteria at different temperatures of incubation can give a direct assessment of its bacteriological quality and of the sanitary conditions under which it was produced. An indication of the size of the active bacterial population of a milk sample can also be obtained indirectly by observing the reduction of coloured dyes, like methylene blue or resazurin, brought about by the respiratory activities of the bacteria.[49]

1 The methylene blue test for milk

Shake the sample of milk to be tested and aseptically add 10 cm^3 to a sterile test-tube. Then add 1 cm^3 of a standard methylene blue solution (see below) with a sterile 1 cm^3 pipette without touching the milk or the test-tube. Insert a sterile rubber bung (sterilised by boiling for at least 5 minutes in distilled water). Invert the tube once or twice to mix and transfer it within 5 minutes to a waterbath fitted with a lid and maintained accurately at 37.5°C $\pm$ 0.5°C.

The level of the water must exceed the level of the milk.

Having set up tests on a series of samples, inspect

the tubes every half-hour, remove tubes which show complete reduction of the blue dye to a colourless form and note the time, those which are *partially* reduced return to bath, and those which show *no* reduction are inverted *once* and then returned.

The methylene blue at the surface of the milk may not be reduced at all—this is ignored.

Two controls need be used:
one, milk and methylene blue as above, but boiled for 3 minutes to destroy the redox system; and the other, milk with 1 cm^3 of water only.

The times taken for the blue colour to disappear may be interpreted as shown in Table 1.

Class of milk	Reduction time	Approximate number of organisms per cm^3
Class I Good	Over 4½ hours	200 000 (or less)
Class II Average	2½ to 4½ hours	200 000—2 000 000
Class III Poor	less than 2½ hours	2 000 000—10 000 000

Table 1 Interpretation of standard methylene blue test

In Britain, the Statutory Instruments (1963 and 1965) require raw and pasteurised milks to pass a series of tests including the methylene blue test. The standard methylene blue solution is prepared for official tests from a commercially prepared tablet made up under Government direction dissolved in 200 cm^3 of cold sterile distilled water under aseptic conditions. The resulting solution is made up to 800 cm^3 and it will keep in a cool dark place for 14 days.

A known volume of a 0.02% aqueous solution of dye to give a pale blue colour may be used to compare different samples though, of course, the figures of Table 1 then no longer apply.

2 The resazurin test for milk

Resazurin solution is used in the same way as methylene blue but it is even more light sensitive and so the test should never be carried out in bright sunlight. The standard solution is prepared by dissolving 0.1 g of resazurin powder in 200 cm^3 of glass distilled water and steaming it for thirty minutes. This *stock* solution should be kept in a refrigerator and diluted 1:10 for use (neither solution will keep for more than 24 hours.) It is important that dilution procedures should be carried out under aseptic conditions. Alternatively, one commercial tablet of resazurin produced under Government direction should be dissolved in 50 cm^3 of sterile water.

After carrying out a test, it is examined by either:

1 noting the colour stage of the reduction after the lapse of the fixed period of time, or
2 noting the time taken for the dye to be completely reduced or reduced to a prescribed colour.

Examination 1 may be used for detecting abnormal milk and 2 for accurate grading of quality milk.

An interpretation of the colour changes which occur are given in Table 2.

Colour	Quality of the milk
Blue Lilac Mauve	Satisfactory
Pink-Mauve Mauve-Pink Pink	Fair
Colourless	Poor

Table 2 Interpretation of standard resazurin test colour after 1 hour of incubation at 37.5°C

Some modification of the test is necessary to take into account the storage temperature of the milk and its effect on the keeping quality, as milk produced in the winter will show less bacterial activity for a given number of organisms than in the summer.

A compensation is therefore made by altering the incubation temperature according to the mean atmospheric shade temperature as shown in Table 3.

Mean atmospheric shade temperature °C	Incubation time in mins. at 37.5°C
4.5 and under	120
4.5–10.0 (inclusive)	90
10.6	75
11.0	65
11.7	55
12.2	50
12.8	45
13.3	40
13.9	35
14.4	30
15.0	25
15.6	20
16.1 and over	15

Table 3 Incubation temperature compensation for resazurin test

The ultimate grading of the milk is given in Table 4.

Grade of milk	Colour	Quality of the milk
Grade I (Category A)	Mauve to Blue	Satisfactory
Grade II (Category B)	Pink-Mauve to Pink	Doubtful
Grade III (Category C)	Pink to colourless	Unsatisfactory

Table 4 Resazurin grading of milk

As a quick test for detecting milk of the poorest quality the resazurin test can be incubated for 10 minutes only and then read.

Except in special circumstances, investigations 3 and 4 should be done as demonstrations only by the teacher.

3 Clostridium perfringens in milk [AL]

This is an anaerobic spore-forming bacterium occurring in the intestine of man and animals as a normal inhabitant. Since it is found in the intestinal content of cows, its presence in milk is indicative of manure contamination from the skin of the animal or the surroundings. Its detection is, therefore, of sanitary value and is based on the reaction known as 'stormy fermentation'.

Take five sterile tubes and place 10 cm^3 of milk to be tested in each. Prepare some sterile paraffin wax.

Figure 12 McCartney bottle with Durham tube

Then:
1 heat the tubes of milk to 80°C for 15 minutes. Cool them in tap water (this eliminates vegetative cells and leaves spores).
2 add melted paraffin wax on top to exclude air (1.5 cm).
3 incubate the tubes at 37°C for 2—5 days.

A 'stormy fermentation' and gas production indicates the presence of *Clostridium perfringens* and other *Clostridium* spp. Should 'stormy fermentation' occur following a test the milk sample should be handled with extreme care BY THE TEACHER ONLY. Because of the nature of the contaminant all positive tubes should be sterilized in an autoclave before disposal of their contents.

Two or more positive tubes out of five are taken to indicate presence of more than 75 mg manure per pint of milk.

4 Presumptive coliform test [AL]

In addition to *Clostridium perfringens* a number of other bacteria occur in the intestines of animals. They include *Escherichia coli* and *Streptococcus faecalis*.

Basically they are all short, gram-negative, non-spore forming rods, which will ferment lactose to produce acid and gas, and, because of their normal ecological niche, they are termed coliform bacteria. If incubated in a broth which includes an indicator (like MacConkey's broth) and an inverted glass tube (Durham tube—see Fig. 12) to collect gas, their presence can be indicated. Bacteria other than coliforms do produce acid and gas and hence the test is only presumptive: it suggests but does not prove the presence of coliforms.

Take:

8 tubes of MacConkey broth, each containing an inverted Durham tube or ignition tube as a substitute and 6 tubes containing 9 cm³ of sterile Ringer's solution as diluent. Three tubes for raw and three for pasteurised milk.

MacConkey broth is prepared by dissolving:

Sodium tauroglycocholate	5 g
Peptone	20 g
Sodium chloride	5 g

in one litre of distilled water. It is best to steam the mixture. Then add 20 g of lactose and adjust the pH to 7.4—7.5. Filter if necessary. Finally, add 10 cm³ of 1% aqueous neutral red and sterilise by steaming for an hour or autoclaving at approximately $69kN/m^2$ for 10 minutes.

The bile salt discourages the growth of non-coliform bacteria.

Ringer's solution is most readily prepared by making a concentrated solution and then diluting this by 3 parts of water before sterilising it for use. The materials for a concentrated solution are:

Sodium chloride (NaCl)	9 g
Potassium chloride (KCl)	0.42 g
Sodium hydrogen carbonate (NaHCO$_3$)	0.2 g
Calcium chloride (CaCl$_2$)	0.48 g
Distilled water	1 litre

Ringer's solution can be made from tablets which are available commercially.

To carry out the test:

1 Dilute both raw and pasteurised milk serially down to 10^{-3} (use a sterile pipette to transfer 1 cm^3 of milk into 9 cm^3 of dilutent. Mix by raising and lowering the liquid six times in and out of the pipette. Rotate the tube between the hands; label 10^{-1} and proceed down to 10^{-3} taking care to use a fresh pipette before each transfer).

2 With one pipette, transfer 1 cm^3 of each dilution starting from the highest dilution, and 1 cm^3 of undiluted milk to each MacConkey Tube.

3 Incubate at 37°C for 48 hours. Then examine for production of acid *and* gas. (Absence of gas from the Durham tube even in presence of acid, i.e. pH indicator turned yellow, is taken as a negative result.)

This presumptive coliform test can also be applied to drinking water and under specified test conditions statistical tables exist to give most probable number (MPN) estimates of the bacterial population.[43,47,51,52]

Table 5 gives an indication of the potability (or drinkability) of water samples.

Presumptive coliforms per 100 cm^3 of sample	Potability Classification
Less than 1	Highly satisfactory
Between 1 and 2	Satisfactory
Between 3 and 10	Suspicious
Greater than 10	Unsatisfactory

Table 5 Potability of drinking water

Foods may be the vehicle of endogenous (e.g. animal diseases such as brucellosis or tuberculosis) or exogenous pathogens. A most important exogenous source of pathogens is man. Intestinal pathogens from human carriers (patients, convalescing persons or chronic carriers) such as typhoid or paratyphoid bacilli are excreted in faeces and urine accompanied by higher numbers of normal intestinal bacteria such as *Escherichia coli*, *Clostridium perfringens* and *Streptococcus faecalis*. Because of their high numbers and as they are easy to detect and quantify, these bacteria (or faecal indicators) are preferred in procedures designed to detect direct or indirect faecal contamination. Their presence, especially that of *E. coli*, means recent faecal pollution and therefore the likelihood of the presence of intestinal pathogens. This is the case especially with water, and some foods such as shellfish and salad vegetables eaten raw, where the demonstration of *E. coli* is alarming at any level in the case of water, or above a certain level in the case of salad foods.

YOGHURT PRODUCTION—A FOOD FERMENTATION PROCESS

This processed food results from a controlled lactic acid fermentation of antibiotic free milk by the symbiotically growing bacteria *Streptococcus thermophilus* (coccus) and *Lactobacillus bulgaricus* (rod). The milk sugar lactose is converted to lactic acid (over 0.65%) and the pH falls from about 6.6 to below 4.6, the isoelectric point of casein. Casein is precipitated. The preservative factor against growth of spoilage of pathogenic bacteria is the lactic acid produced, although the low pH does not necessarily free yoghurt from human pathogens, which may already be present before processing starts, neither does it protect against spoilage by yeasts or moulds. It is therefore essential that high grade raw milk free from antibiotics and other microbial inhibitors (e.g. sanitisers) is used and rendered safe from possible indigenous pathogens (e.g. tubercle bacillus) by appropriate heating.

Take:

a starter culture (unpasteurised commercial yoghurt diluted ten times)
raw milk; 50 cm^3 in a 100 cm^3 conical flask

BDH pH indicators:	Bromo-phenol blue	(pH 2.8—4.4)
	Methyl-red	(pH 4.4—6.0)
	Bromo-cresol purple	(pH 5.8—6.8)

Then:
1 heat the raw milk to $80^{\circ}C$ for 30 minutes.
2 cool it in tap water and adjust temperature to $45^{\circ}C$.
3 pipette $2\ cm^3$ of starter, mix and record the pH by transferring a drop of milk to a white tile with a pasteur pipette. Add an equal amount of indicator and compare the colour produced with the standards for the indicators.
4 place the mixture in the $45^{\circ}C$ waterbath and leave it for three hours. Take the pH every thirty minutes.

It is interesting to make a graphical plot of pH against time.

MICROBIAL SPOILAGE

Demonstrations of the type of reaction which occur are:

1 Protein breakdown or proteolysis

Take one milk agar plate prepared by adding aseptically $1\ cm^3$ of sterile reconstituted skim milk (10% weight to volume of solid to water) to $10\ cm^3$ of nutrient agar in a Petri dish. Mix by shaking gently on a bench surface and allow it to set. Then dividing the plate into two sectors, mark and streak one sector with *Escherichia coli* and one with *Chromobacterium lividum*. Incubate at $20-25^{\circ}C$ for about two days and examine. A clear zone of casein breakdown will develop around the *Chromobacterium* but not around the *Escherichia*.

2 Lipolysis

Fat-splitting organisms cause spoilage in butter, margarine and lard. Most of them will split glyceryl tributyrate (tributyrin) and so this can be incorporated into a test medium.

Prepare a tributyrin plate from the following medium:

Peptone	0.5 g
Yeast extract	0.3 g
Tributyrin	1.0 g
Agar	1.2 g
Distilled water	$100\ cm^3$

Then using the same procedure as used to show proteolysis, streak the plate and incubate with bacteria. A clear zone developed around a growth indicates lipolysis. Most proteolytic bacteria are also lipolytic.

28

3 Pectolysis

Soft-rot bacteria dissolve the middle lamellae of host cells of plants, fruits and vegetables, causing the tissues to disintegrate. This is achieved through the ability of the pathogen to secrete pectolytic enzymes which break down the pectins in the middle lamellae and other cell parts.

The pectolysis carried out by *Erwinia atroseptica*, a bacterium responsible for the serious disease of potatoes known as Blackleg, can be demonstrated as follows:

Place turgid discs of potato tuber tissue 8.0 mm. x 0.5 mm. thick in damp chambers consisting of Petri dishes lined with moist filter papers. To each disc add a loopful of *E. atroseptica* cells taken from a slope culture and incubate at $25^{\circ}C$ overnight. Soft rotting is easily seen the next day.

4 Starch breakdown (amylase production)

Take nutrient agar slope cultures of:

Escherichia coli ⎫
Bacillus cereus ⎬ bacteria

Saccharomyces cerevisiae — yeast

and a starch agar plate prepared by mixing $9\ cm^3$ of sterile molten nutrient agar with $1\ cm^3$ of sterile 2% starch paste. When the plate has set streak sectors of it with the organisms and incubate it at $30^{\circ}C$ for 2 days. Then treat the surface with dilute iodine solution and drain it. Hydrolysis of starch is shown by a clear zone. It is interesting to note that the brewer's yeast is devoid of amylase and hence the necessity of the 'mashing process' (early germination and soaking of the barley grains) in beer manufacture.

SPOILAGE OF CANNED FOOD [AL]

Many canned foods are spoiled by bacteria which are able to withstand high temperatures (thermophiles). Flat-sour spoilage is caused by non gas-producing thermophiles and 'Thermophilic Anaerobe' or 'TA' spoilage is caused by those which are gas-producing.

For a worthwhile demonstration, prepare a nutrient broth culture of *Bacillus stearothermophilus* propagated aerobically and a liver broth culture of *Clostridium thermosaccharolyticum* propagated anaerobically. Both cultures should be incubated at $55^{\circ}C$.

Liver broth is most conveniently made from commercially prepared granules. It can however be made by soaking about 500 g of fresh liver pieces for

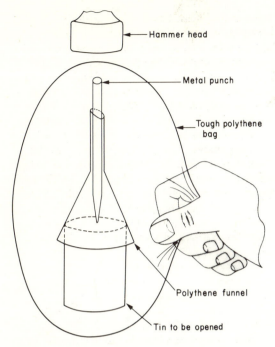

Figure 13 Funnel with metal punch for opening cans

24 hours in water, then heating to 80°C and straining off the juice through muslin, mixing the liquid with 5 g of sodium chloride and 10 g of peptone before making the total volume up to one litre with distilled water.

Take: 3 cans of corn (maize), a can opener, solder and soldering iron.

Then:

1 flame the top of two of the cans.
2 open a small hole (0.25 cm wide) on each of the two cans.
3 by means of a Pasteur pipette, add one cm³ of *B. stearothermophilus* to one can and about 5 cm³ of *C. thermosaccharolyticum* to the second. The third can is the control.
4 seal the opened cans with solder.
5 place all cans in tough plastic bags, twisted at the end and fastened with rubber bands.
6 incubate all three cans at 55°C for 2–5 days, but not more.

The can treated with *C. thermosaccharolyticum* will develop a bulge. Open all three cans, record the pH of each and note the colour and appearance of the corn.

Great care is needed when opening a swollen can. It is best tackled with a metal punch placed in the stem of a plastic funnel which is fitted over the can (see Fig. 13) and with the whole of the equipment inside a large tough plastic bag.

Even more care is needed if the cause of a 'swell' is not known and the can is to be opened. If for instance it is due to *Clostridium botulinum* pre-formed toxin may reach the conjunctiva, with scattered food particles, and produce botulism. Commercial cans which are found to be swollen should NEVER be opened in school laboratories.

SPOILAGE OF PRE-COOKED FOODS

Pre-cooked foods, like sliced ham, are particularly liable to microbial contamination and hence spoilage. To estimate the extent of the contamination of a cooked meat, an assay is best performed in which the organisms are 'washed' free from the food substrate and allowed to develop. In the teaching situation, the procedure can be demonstrated by placing small pieces of freshly cut ham in three empty sterile Petri dishes and replacing the lids.

Then one dish containing ham is stored in a refrigerator for 2 days, the second is kept in a warm room for 2 days and the third is prepared immediately for assay by the addition of about 15 cm³ of melted nutrient agar medium which has been cooled to 45–50°C. The dish is swilled very gently as the agar is poured onto the ham, and the lid replaced. The rotary movement removes bacteria from the surface of the meat which itself sinks and, after incubation, colonies of bacteria can be seen growing on the agar medium above the samples.

The two ham dishes stored at 4°C and room temperature are assayed after 2 days using the same procedure. The agar is allowed to set in all the plates before incubating them at 37.5°C for 1 or 2 days.

Cultures can be prepared from other foods by covering them with agar using this procedure too.

SELECTIVE CULTURE OF BACTERIA [AL]

Natural selection of a particular organism at the expense of others is an important principle in ecology and evolution and artificial selection has had a considerable impact in clinical bacteriology and in microbial genetics.

To select an organism in artificial culture, the conditions are arranged to favour its multiplication and discourage the growth of others also present in the inoculum.

A good garden soil contains a variety of bacteria and by producing a basic mineral medium and then varying:

1 the energy and carbon sources available,
2 the nitrogen source available, and
3 the terminal hydrogen acceptor available for respiration, it is possible to select a range of bacteria.

Given a basic medium the conditions which favour particular bacteria are given in Table 6.

Bacteria	Energy Source	Nitrogen Source	Hydrogen Acceptor
Azotobacter	Alcohol	Nitrogen gas	Oxygen gas
Aerobic bacteria	Alcohol	Ammonium chloride	Oxygen gas
Denitrifying bacteria	Alcohol	Sodium nitrate	Oxygen gas or nitrate ion
Desulpho vibrio *	Alcohol	Ammonium chloride	Sulphate ions
Clostridium pasteurianum	Glucose	Nitrogen gas	Organic compounds derived from glucose eg pyruvic acid.

Table 6 Conditions for the selective culture of soil bacteria

* The Desulphovibrios are sulphate reducing organisms. They cause the black deposits and smell of sulphide just below the surface of sands and muds of rivers and the sea shore. They are unlikely to be present in cultivated garden soil and so if it is desired to culture them it is necessary to add a little pond mud to the inoculum soil used.

A convenient demonstration can be done by preparing the following solutions:

1 Mineral salts.

Magnesium sulphate ($MgSO_4.7H_2O$)	0.2 g
Di-potassium hydrogen orthophosphate (K_2HPO_4)	1.0 g
Iron (II) sulphate ($FeSO_4.7H_2O$)	0.05 g
Calcium chloride ($CaCl_2$)	0.02 g
Manganous chloride ($MnCl_2.4H_2O$ (or sulphate)	0.002 g
Sodium molybdate ($Na_2MoO_4.2H_2O$)	0.001 g
Distilled water	330 cm^3

2 Absolute alcohol (ethanol) 2.5 cm^3 in 15 cm^3 distilled water.

3 Glucose 2 g in 40 cm^3 distilled water.

4 Ammonium chloride 0.5 g in 100 cm^3 distilled water.

5 Sodium nitrate 1.0 g in 65 cm^3 distilled water.

6 Sodium sulphate 1.0 g in 65 cm^3 distilled water.

Mix the solutions as shown in Table 7 to give 30 cm^3 aliquots of culture medium. 100 or 150 cm^3 conical flasks are convenient culture vessels except for *Desulphovibrio* and *Clostridium* culture. For *Desulphovibrio* use about a 28.4cm^3 (1oz) screw cap bottle and for *Clostridium* use about a 142 cm^3 (5oz) screw cap bottle. It is unnecessary to sterilise the media if they are to be used immediately. The screw cap bottles should be heated in a water bath to displace oxygen from the culture medium and then the one for *Desulphovibrio* should be filled completely with freshly boiled water. Screw the caps up tightly and allow to cool. Inoculate each culture medium with a few milligrams of garden soil or other material. An inoculum in excess of this will contain nutrients in quantities which will destroy the selective effects of the synthetic medium. After inoculation the *Clostridium* bottle should be flushed with nitrogen gas by bubbling for a few minutes and then securing the cap as rapidly as possible.

The cultures are best incubated between $25°-30°C$.

Solution	1	2	3	4	5	6	Distilled water
Azotobacter	10	1.0					19
Aerobic bacteria	10	1.0	6				13
Denitrifying bacteria	10	1.0		6			13
Desulpho-vibrio	10	1.0	6		10		*
Clostridium pasteurianum	10		6				+14

Note * Fill culture container with boiled water.
   + Add 0.15 g of chalk (Ca CO$_3$) to each culture.

Table 7 Mixing solutions for culture medium, cm^3 of solutions for 30 cm^3 of culture medium

It is not possible to give precise times at which organisms will appear but the following times are typical. The cultures will never be entirely of one species of organism but one type will certainly predominate.

Azotobacter

In 4—7 days a surface film (pellicle) indicates *Azotobacter*. Examine a smear microscopically using capsule stain (see below). The cells are elliptical and large (about the size of yeast cells), usually in pairs and enclosed in a slimy capsule. Further purification can be made on nitrogen free agar medium. The cultures, particularly after purification, can be used to show the effect of aeration versus no aeration in media with pH values of 5.5, 7 and 8.5.

Aerobic bacteria

After 2 days, as the concentration of bacteria rises above 10^6 cm^{-3} both gram-positive and gram-negative cells are produced but gram-negative organisms predominate.

Denitrifying bacteria

After 2 days, an even turbidity is seen. The bacteria are mostly gram-negative and can be subcultured to a nitrate broth of 1% potassium nitrate and 0.01% peptone with a Durham tube to collect nitrogen gas.

Desulphovibrio

After some 3—4 weeks or more an even turbidity with black particles and sediment develops. There is a smell of hydrogen sulphide. The bacteria are gram-negative rods.

Clostridium pasteurianum

After 4—6 days, an even turbidity is seen. Gram staining shows long slender positive rods. With a spore stain (see below), large terminal spores (drumstick type) are revealed. Some purification of any of these species can be made by subculture into fresh medium but sometimes this is unsuccessful because of the absence of trace amounts of nutrients included in the original soil inoculum.

A suitable capsule staining method is to mix a loopful of Indian ink with a loopful of culture in 5% dextrose solution at one end of a microscope slide. Spread the mixture with the end of another slide as in the preparation of a blood film. Allow the smear to dry and then pour a little methyl alcohol over it to fix it. Stain for a few seconds with crystal violet (see page 18). The organisms appear stained blue with the capsules showing as haloes.

For spore staining[4],[6] make a thick film and stain for 3 minutes with hot carbol fuchsin (see page 18). Wash and then flood with 30% aqueous ferric chloride for 2 minutes. Decolourise with 5% sodium sulphate solution, wash, and counterstain with 1% aqueous malachite green. The cells are stained green and the spores red.

Some of the autotrophic bacteria, like the sulphur bacteria which develop in a Winogradsky column, are particularly sensitive to the nature of their surroundings, but they can be successfully propagated with an appropriate enrichment culture medium made from the following:

1 Aerobic sulphur oxidising bacteria (*Thiobacillus*)
Requirements
A basal medium made by dissolving—

Potassium dihydrogen orthophosphate (KH$_2$PO$_4$)		4 g
Dipotassium hydrogen orthophosphate (K$_2$HPO$_4$)		4 g
Ammonium chloride (NH$_4$Cl)	0.5 g	0.5 g
Hydrated magnesium sulphate (Mg SO$_4$.7H$_2$O)		0.8 g
Sodium chloride (NaCl)		0.25 g
Calcium chloride (CaCl$_2$)		0.1 g

Iron (II) sulphate ($FeSO_4.7H_2O$)   0.01 g
in one litre of distilled water.

Procedure

Pipette 25 cm^3 of medium into each of three 250 cm^3 conical flasks. To two flasks add 3 cm^3 of 10% sodium thiosulphate ($Na_2S_2O_3.5H_2O$), to the third flask add 3 cm^3 of water. Inoculate one of the thiosulphate flasks and the flask without thiosulphate with 2 cm^3 of canal water or 2—3 g of soil. Do not inoculate the second thiosulphate flask. Incubate at 30°C for 1—2 weeks. Then examine cultures for surface pellicle formation and production of elemental sulphur, acid production by determining the pH, and the presence of straight gram-negative rods.

2   Anaerobic, photosynthetic sulphur oxidising bacteria *(Chlorobium, Chromatium)*

Requirements

A basal medium made by mixing—

Sodium hydrogen carbonate ($NaHCO_3$)	2 g
Sodium chloride ($NaCl$)	0.25 g
Calcium chloride ($CaCl_2$)	0.1 g
Iron II sulphate ($FeSO_4.7H_2O$)	0.1 g

in one litre of distilled water.

Procedure

Add 1 cm^3 of 20% sodium sulphide to 200 cm^3 of the basal medium. Then add 1 cm^3 0.1 M hydrochloric acid and mix making the addition in a fume cupboard. Dispense the medium into small reagent bottles (about 60 cm^3) having ground glass stoppers. Inoculate one bottle with 1—2 g soil and another with 1—2 g river mud. *Completely* fill the bottles with medium and seal with the glass stoppers. Incubate in the light for 1—2 weeks at room temperature and observe for the development of green and purple bacteria. The green bacteria deposit sulphur externally, the purple bacteria internally.

3   Anaerobic photosynthetic, non-sulphur bacteria *(Rhodospirillum, Rhodopseudomonas)*

Requirements

The same basal medium as for the other anaerobic bacteria above.

Procedure

Add 1 cm^3 of 10% sodium malate solution to a 60 cm^3 reagent bottle and half fill the bottle with basal medium. Inoculate with river mud (1—2 g), completely fill the bottle and stopper. Incubate in the light for 1—2 weeks at room temperature.

Rhodospirillum species are gram-negative, rod-shaped or spherical and the colony colour varies from yellowish-brown to reddish-brown, depending on the conditions of light and oxygen concentration.

ESTIMATING NUMBERS OF MICROBES [AL]

To estimate the success of an enrichment culture or assess the rate of growth, or size, of a population of micro-organisms a variety of techniques can be used.

Viable counts

These are methods aimed at estimating the number of bacteria capable of *growth and multiplication* in the material under examination. Samples of the material must be collected in sterile containers, with precaution against contamination, and aseptic procedures used throughout. They are then diluted, inoculated onto a medium and the colonies which develop from single organisms counted.

1   The preparation of dilutions

This is the first essential step for all methods, since the original material will frequently contain too many organisms to count without dilution. If the material under study is a solid, a weighed amount must first be suspended in a known volume of sterile diluent, well mixed by shaking, and then further dilutions made from this. If the material is a liquid, dilutions are made from it directly, after thorough shaking which is again necessary to ensure an even distribution of the bacteria.

For most general purposes sterile Ringer's solution (see page 27) is ideal. Physiological saline of 0.85% sodium chloride only is not suitable as it often causes cells to clump.

Measured quantities of sterile diluent are placed in sterile tubes or bottles by means of sterile pipettes or some other special device for aseptic distribution: for example, a burette fed through a slide-arm from an overhead reservoir of sterile diluent, and provided with a protective hood over the jet. Unfortunately, measured volumes cannot be sterilised by heat without danger of changing the volume.

A ten-fold serial dilution for viable counts are made as follows:

1   prepare the required number of tubes containing 9 cm^3 sterile dilutent.

2   with a sterile 1 cm^3 blow-out pipette fitted with a teat, add 1 cm^3 of the well shaken sample to one container and 9 cm^3 of sterile dilutent. *Discard*

this pipette.

3 with a fresh pipette, draw the liquid up and down ten times to ensure thorough mixing and then add 1 cm^3 of this 1/10 dilution to another 9 cm^3 of dilutent.

4 again discard the pipette, and use a third for mixing this dilution and making the next.

5 repeat this procedure until you reach the required degree of dilution, but avoid mixing the last dilution until you are ready to use it.

The number of dilutions needed will depend on the material concerned, only the two or three greatest being actually used (except sometimes in preliminary experiments.) To investigate tap water, dilutions of 1/10 and 1/100 are normally needed whilst for pasteurised milk 1/100 and 1/10 000 are better. For water suspected of pollution, dilutions up to 1/10 000 and for soil up to 1/100 000 are usually suitable. A visibly turbid broth culture of bacteria contains about 10^8 to 10^9 cells per cm^3.

2 The dilution plate method

This is the most widely used procedure. Ten-fold dilutions are prepared as described and plated out as follows:

1 with the pipette used for *mixing* the last dilution place 1 cm^3 of this dilution in each of two or three sterile Petri dishes, then 1 cm^3 of the next (lower) dilution in each of two or more sterile dishes, and so on for as many dilutions as are to be plated.

It is seldom necessary to plate more than two or three dilutions, but even if a long series is to be plated, the same pipette can be used for the whole series provided that the dilutions are taken in the right order (high to low) and that the pipette is not allowed to become contaminated. Each dilution used must be plated in duplicate and, if time and material allow, greater replication is desirable even in routine work.

2 when all dilutions have been plated out, add to each plate 15 or 20 cm^3 of a suitable agar medium, previously melted and cooled, and make sure that the sample is evenly distributed before the agar sets.

3 incubate the plates at a suitable temperature. After incubation discard any plates showing more than 300 or fewer than 30 colonies. Errors due to overcrowding and errors of sampling make counts outside these limits unreliable. Thus, if ten-fold

dilutions are used, probably only one set of plates will actually be suitable for counting, and dilutions should be selected for plating with this in mind.

Mark the plates with wax pencil, on the under surface in sectors or in a grid pattern, making *thin* lines, and count all the colonies on each plate of the appropriate dilution. Take the mean of the count for the two (or more) plates, multiply by the dilution factor, and express the result as the 'Number of colonies per cm^3 or g of the original material, developing at . . °C in . . days on . . . medium' rather than as the 'Number of bacteria per ml or gm of the original material'. It will often be found useful to use tally counters for counting.

Strictly speaking, the common assumption that one colony represents one bacterium in the original material is not justified, even when the material is a suspension of a single species. For example, if the bacteria form clumps or chains, each such group will give rise to a single colony unless it is broken up during dilution.

A further problem which occurs is that if the material contains a mixed microbial flora, it is extremely improbable that all the species present will develop equally well on the same medium or under the same conditions of incubation.

3 The spread plate method

In this method the dried surface of suitable agar plates is inoculated with prepared dilutions. From the appropriate dilutions, 0.1 cm^3 of the cell suspension is pipetted aseptically onto the surface of each of three plates. A sterile glass spreader (see Fig. 14) is used to spread the organisms over the surface of the plate. After sterilisation the spreader must be cooled before use. This can be done by touching it on the edge of the plate. After allowing the inoculated liquid to soak into the agar, the plates are incubated until single cells develop into colonies which can be counted.

Total counts

This covers methods of estimating total number of organisms per unit volume or weight of sample without distinguishing between viable and non-viable cells.

1 Using a haemocytometer

The actual number of organisms can be determined by counting in a suitable chamber like the

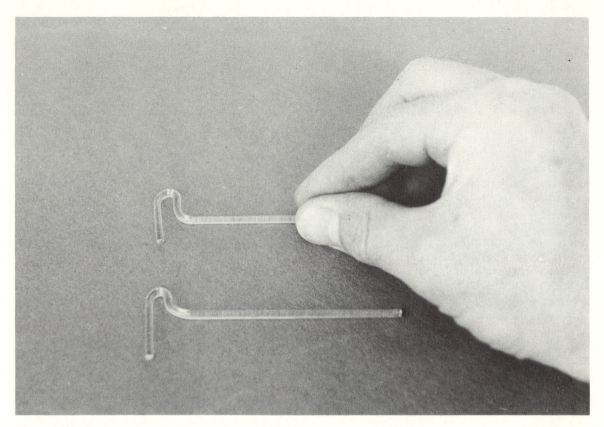

Figure 14 A glass spreader

Fuchs—Rosenthal ruled haemocytometer (See Fig. 15). Appropriate dilutions are prepared, though sterile dilutent is not necessary, to give at least 400 organisms per slide.

To use a haemocytometer slide:

1　examine the slide with a hand lens and see where the graticule grid is.
2　clean the slide and cover glass with a dry tissue or cloth.
3　breathe on the slide to moisten it very slightly and then put the cover glass in position over the graticule by sliding it slowly on to the slide from the edge nearest to you. Keep it flat. Use both index fingers on the cover glass, pressing downwards firmly, but only where it is supported by the slide underneath. Move the cover glass across the slide by using both thumbs. One edge of the cover glass should project above the outer, vertical, edge of the slide.
4　make sure the cover glass is correctly attached to the slide by checking that rainbow-like patterns (Newton's rings) are visible over most of the area of contact. Repeat 3 if necessary. The junction between slide and cover glass should be sufficiently strong to keep the slide clinging to the cover glass when it is lifted.
5　then use a fine pipette or syringe with a wide diameter needle to introduce the microbial suspension as a drop below the edge of the cover glass, filling the space between the platform and the cover glass almost completely. Do not over- or under-fill. There should not be any suspension in the moats around the platform.
6　remove any surplus fluid from the top, sides, and

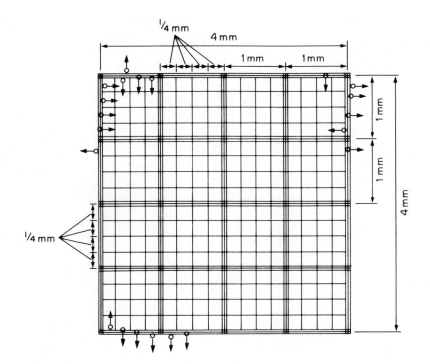

Figure 15 Rulings on the base of the counting chamber of a Fuchs-Rosenthal haemocytometer. Cells touching any of the triple boundary lines along two sides at right angles are counted. Those touching the remaining two sides of a marked square are not counted. The symbol → is used to indicate which cells should be counted. It is simplest to start counting at the top left hand square of any block as seen by the observer. (Diagram reproduced from Bradbury, S., Peacock's Elementary Microtechniques, Fig. 7.5, Edward Arnold (Publishers) Ltd., 1973).

bottom of the slide and cover glass with filter paper.

7 organisms are found and counted under the medium power of the microscope. Some cells will lie on a boundary between squares. The usual procedure is to count the cells on the north and west boundaries and to ignore those on the south and east. It does not matter whether a cell is mostly in a neighbouring square; count it if it touches the north or west boundary. Otherwise ignore it.

The volume of suspension over the rulings can be found from the size information given on the slide.

2 Smear count

A known volume of the source material, diluted as necessary, is spread evenly over a defined area on a clean slide. The film is stained by a suitable method and the average number of organisms in a field of view determined by counting in several fields chosen at random. The area of the microscopic field is determined, and hence the number of organisms per cm^3 of the source sample can be calculated.

OPTICAL DENSITY MEASUREMENTS [AL]

Measurements of optical density or turbidity are convenient to follow changes in population size of micro-organisms and can be used to study growth curves. There are two main types of electrical apparatus which enable readings to be obtained: both make use of a photo-electric cell (which generates current) or a photo-resistor (which varies an applied current) depending upon the amount of light received.

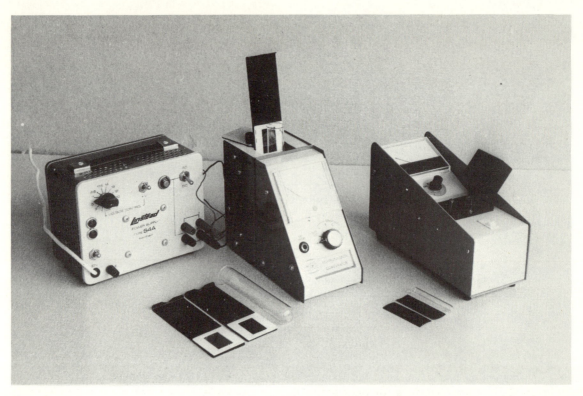

Figure 16 Some colorimeters and a nephelometer. (a) Two typical colorimeters of the type used for elementary studies. In each of them the light source, photo-electric cell, sample tube and meter are in the same case.

In the *colorimeter* or *absorptiometer* type (Fig. 16a) the sample, in a special container, is placed between a source of light and the photo-sensitive unit which is only exposed to a direct beam of light passing through the sample. The current from the unit will depend on the intensity of the light reaching it and thus the greater the turbidity (or depth of colour) of the suspension, the smaller will be the current. This is recorded on a scale so that turbidity readings can be obtained for any number of samples and compared with counts obtained by other means. It is important to realise that measurements of turbidity have no meaning in themselves and must always be translated into terms of either bacterial dry weight, of cell concentration, or of viable counts. The apparatus must be calibrated *for each bacterial species used* by plotting the turbidity readings obtained for a series of dilutions against numerical

values obtained for the same dilutions by a viable or total count procedure. By using the calibration graph so obtained, turbidity readings obtained for further suspensions of the same species can be converted into actual values.

In the second type (*nephelometer*, see Fig. 16b) the sample in a special tube is placed in a socket in the instrument and illuminated from below. A cover is placed over the tube before the light is switched on. An iris diaphragm controls the intensity of the beam which is focused by the hemispherical base of the tube acting as a lens and passes directly up the centre of the tube. If the liquid is turbid, light is scattered by the suspended particles. The base of the tube is surrounded by a ring of electro-sensitive units inclined at an angle so that they receive no light except that scattered from the particles in the tube by reflection. As in the absorptiometer, the current

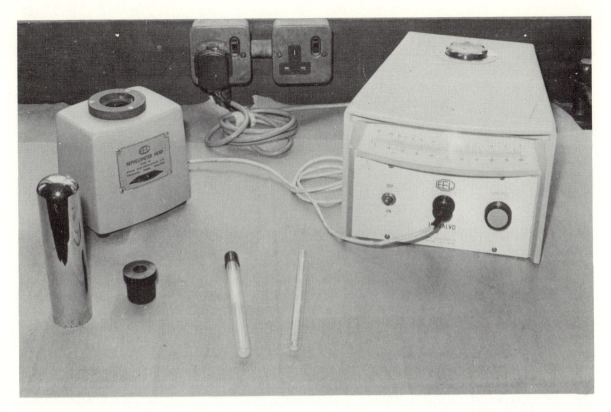

Figure 16(b) A nephelometer with the electro-sensitive unit on the left and a galvanometer to read the current produced as a result of light scattered from particles in fluids placed in the sample tubes. The special sample tubes are in the centre of the picture together with the cover which is placed over them before taking readings.

generated in the units (which here will be the greater the more turbid the sample) is recorded on a meter.

DETERMINATION OF CELL DRY WEIGHT [AL]

The dry weight of bacterial substance per cm^3 of a suspension can be measured directly provided that the suspension contains no foreign matter.

Place two or more carefully measured samples in accurately weighed watch-glasses. Allow the samples to evaporate to dryness at about $50°-60°C$ and determine their dry weight. From the mean of the results, the bacterial density can be calculated. If the bacteria are suspended in any fluid other than distilled water, the dry weight per cm^3 of this fluid must be determined and subtracted from the figure obtained for the suspension as a whole.

The degree of accuracy that can be obtained by

this method depends on the skill and care exercised and the reliability of the apparatus used. The larger the mass of bacteria used the greater the accuracy.

THE BACTERIAL GROWTH CURVE [AL]

The growth of an inoculum of *Escherichia coli* in nutrient broth can be readily followed by viable count procedures, by turbidity measurements or by isotope tracer technique.[36] However, growth can only be followed optically when the population size reaches about 2×10^7 cells per cm^3 as concentrations of cells less than this do not render the medium opaque.

Typical strains of *E. coli* divide approximately every twenty minutes in nutrient broth at $37°C$. If an inoculum from a stationary phase (overnight) culture is used there is a lag of about an hour, assuming that the new medium is equilibrated to $37°C$. During the

37

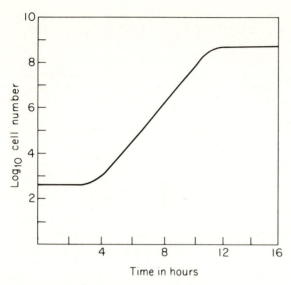

Figure 17 Growth curve of E. coli

taken at 30 minute intervals and dilutions carried out as shown in Table 8.

Time in minutes	Dilutions to be plated in duplicate		
0 (Inoculation)	10^{-2}	10^{-3}	10^{-4}
30	10^{-2}	10^{-3}	10^{-4}
60	10^{-2}	10^{-3}	10^{-4}
90	10^{-3}	10^{-4}	10^{-5}
120	10^{-3}	10^{-4}	10^{-5}
150	10^{-4}	10^{-5}	10^{-6}
180	10^{-4}	10^{-5}	10^{-6}
210	10^{-5}	10^{-6}	10^{-7}
240	10^{-5}	10^{-6}	10^{-7}
270	10^{-6}	10^{-7}	10^{-8}
300	10^{-7}	10^{-7}	10^{-8}
330	10^{-7}	10^{-8}	10^{-9}
360	10^{-7}	10^{-8}	10^{-9}
390	10^{-7}	10^{-8}	10^{-9}

Table 8 Sample times and plating dilutions to follow the growth of a population of E. coli at $37^{\circ}C$

lag phase the viable population remains the same (see Fig. 17). For the next half hour or so the growth rate accelerates to the maximum when the cells are growing exponentially. If an inoculum of 10^5 cells per cm^3 of culture medium is used then dilutions of 10^{-2} 10^{-3} and 10^{-4} are plated out. For example, a cm^3 sample of a 10^{-3} dilution should yield 100 colonies which is a convenient and statistically satisfactory number of colonies to count if the samples are plated in duplicate. As the culture grows the dilutions plated are adjusted. After a time essential components of the medium will become exhausted and become growth limiting when the population reaches about 5×10^8 cells per cm^3. Growth will cease when the culture reaches about 4×10^9 cells per cm^3. In static cultures oxygen becomes limiting before any of the medium components and growth will slow down sooner in an unagitated medium at a population of about 10^8 cells per cm^3.

A convenient practical procedure when plating for viable counting is to take a 2 litre flask containing 200 cm^3 of nutrient broth which is warmed to $37^{\circ}C$. It should be mechanically stirred or agitated and then inoculated with 0.01 cm^3 of an overnight broth culture (or 1.0 cm^3 of a 10^{-1} dilution in warm broth to avoid temperature shock.) This gives an initial population of 10^5 cells per cm^3. Samples for plating are

For absorptiometer or nephelometer measurement, cultures are set up in the tubes or flasks used in the instruments. From the data obtained estimates of specific growth rates, doubling time, initial viable population, final stationary phase population, length of lag, and length of exponential phase can be made.

The relatively long lag phase and moderate growth rate of *E. coli* means that this investigation often cannot be fitted conveniently into the school timetable. It has been shown recently that *Vibrio natriegens* is particularly suitable for investigations of growth curves. Full details of the technique may be found in reference 45.

DIAUXIC GROWTH [AL]

Diauxic growth is often observed when a bacterial culture is grown in the presence of two carbon sources. It is characterised by two phases of exponential growth, separated by a transient lag phase. During the first exponential phase one of the carbon sources is utilised preferentially, while utilisation of the other carbon source is repressed. This repression is brought about directly by the first carbon source, or by its metabolic products. It is due primarily to inhibition of the synthesis of enzymes and permeases involved in the utilisation of the second carbon source. As the first carbon source inhibits the synthesis of these compounds required for the utilisation of the second carbon source, when the first carbon source is exhausted a transient lag phase may follow while the enzymes and permeases required for the utilisation of the second carbon source are synthesised.

In *E. coli* glucose inhibits the synthesis of many of the enzymes involved in the utilisation of other carbon sources. It inhibits the utilisation of glycerol by repression of the synthesis of glycerol permease, glycerol kinase and L-α-glycerophosphate dehydrogenase and so diauxic growth occurs if *E. coli* is inoculated into a mixed medium of mineral salts, glucose and glycerol. A good basic mineral salt medium is:

Disodium hydrogen orthophosphate ($Na_2HPO_4.2H_2O$)	7.8 g/l
Di-potassium hydrogen orthophosphate (K_2HPO_4)	8.4 g/l
Ammonium sulphate ((NH_4)$_2SO_4$)	2.0 g/l
Magnesium chloride ($MgCl_2$)	0.13 g/l
Iron II sulphate ($FeSO_4.7H_2O$)	7 mg/l
Manganous chloride ($MnCl_2.4H_2O$)	1 mg/l
Zinc oxide (ZnO)	0.9 mg/l
Cobalt chloride ($CoCl_2.6H_2O$)	0.2 mg/l
Copper II chloride ($CuCl_2.2H_2O$)	0.3 mg/l
Boric acid (H_3BO_3)	0.1 mg/l
Sodium molybdate ($Na_2MoO_4.2H_2O$)	0.3 mg/l

To carry out a comparative investigation load three suitable containers at 37°C with 20 cm^3 aliquots of mineral medium and add carbon sources as in Table 9.

Aqueous solutions	Volume of liquid (cm^3) to be placed in containers		
	1	2	3
Glucose (2%)	2.0	0.5	Nil
Glycerol (5%)	Nil	1.5	2.0

Table 9 Carbon sources for diauxic growth of E. coli

Provided that the initial inoculation with *E. coli* (preferably growing in the glucose-mineral salts medium at 37°C) gives a concentration of the order of 10^5 bacteria per cm^3 the growth can be observed by following changes in optical density. Take readings at 30 minute intervals and incubate the container with agitation at 37°C. The growth rates on the two substrates (glucose and glycerol) and the length of the diauxic lag are most readily obtained by interpretation of a plot of the log of the optical density against time.

TEACHING METHOD

It is possible to use bacteria to illustrate a wide range of fundamental biological processes. Because of their ubiquity and small size, the organisms themselves have an inherent fascination and they can be manipulated quite simply. At the same time however, as is readily seen in the increasing complexity of the manipulations in this chapter, bacteriological investigations can require considerable technical skill and a little mathematics to cope with serial dilutions, calibration curves and the statistics of probable number estimates. In the field of general biological education some work on, and with, bacteria should be included but only the individual teacher can really select those items most suited to his, or her, students.

Chapter 3 Mycological investigations

Fungal spores are widely distributed in the atmosphere and they will readily develop on a range of raw produce, foodstuffs and a surprising variety of manufactured goods. Telegraph poles and fence posts are destroyed by wood rotting fungi; cotton cloth and papers composed of cellulose are quickly spoiled by fungi which produce cellulases, and even electrical apparatus can be put out of action by hyphal growths which cause short circuits. A host of fungal parasites ravage crops and others cause diseases of livestock and man. In temperate zones human mycoses (fungus caused diseases) like 'thrush' and 'ringworm' were generally considered to be of minor importance but in recent years more serious diseases, like 'farmer's lung', have been diagnosed which match many of the endemic fungal diseases of the tropical countries in their severity. However, in ecological systems fungi do play an essential part in decay and the recycling of materials. Students often fail to appreciate the rapidly growing number of microfungi which are extremely useful to man in the production of valuable organic compounds as well as antibiotics.

The fungi, and in particular the microfungi, are a fascinating group of organisms. The range of structures and life histories which they exhibit are very wide and with a little patience it is possible to present students with an instructive and beautiful range of living colonies which will forever dispel the drab image of mycology so readily developed by rigid studies of representative types.[3,12,17,20,50]

SIMPLE METHODS FOR DEMONSTRATING THE PRESENCE OF FUNGI IN THE AIR

1 Spores in the air
Leave plates of malt agar open to the air for 20—30 minutes at various localities. During this time, spores will settle on to the plates and, if the plates are subsequently covered and incubated at room temperature for a few days, hyphal colonies may be observed.

2 Spores from decaying leaves
The range of fungi in the leaf litter of natural woodland is fascinating and a useful method for

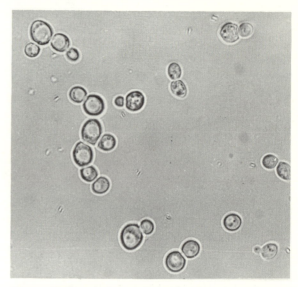

Figure 18 Photomicrograph of budding yeast cells. Unstained cells x 590 showing multipolar budding

cultivating some of them is to fix fallen leaves to the inside of the lid of a Petri dish of malt agar with Sellotape. Left with the lid uppermost spores fall from the leaves on to the agar where colonies develop free from plant detritus. The fungi which can be recovered from coniferous needles is very different from that which can be isolated from broad-leaved species and the comparisons which can be made are most relevant to the ecological differences in the two woodland types.

3 From mouldy substrates
Take a flamed and cooled loop, touch it against the mould growth to be investigated and then streak it across a malt agar plate. The streaking will tend to separate spores out so that discrete colonies develop and it is possible to make a reliable estimate of the number of species actually growing in the substrate.

4 Isolation from soil
Put a drop of sterile water in a Petri dish and

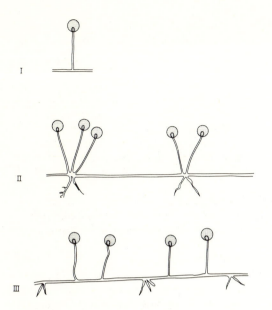

Figure 19 Some common representatives of the order Mucorales

inoculate it with a small amount of soil. Prepare malt agar containing Rose Bengal, a bacteriostatic dye (2.0 g of malt extract, 1.5 g agar, 0.006 g Rose Bengal and 100 cm³ of water) and allow it to cool until it is a comfortable temperature to the hand. Add the agar to the Petri dish moving it gently to distribute the soil, and allow it to set. Store the plate for about a week and then examine it. The Rose Bengal will limit the growth of some of the faster growing fungi so that the plate does not become swamped by one rampant species.

GENERAL CHARACTERISTICS OF ORGANISMS WHICH MAY DEVELOP ON EXPOSED PLATES AND THOSE TREATED WITH GROSS INOCULA

In addition to the typical thread masses, so characteristic of a fungal mycelium, slimy colonies with no obvious structure may occur. If the cells of such a colony are small then the organism involved is likely to be a bacterium, but if the colonies are cream or pink with large cells showing buds (Fig. 18) the organism is a yeast.

Fast growing mycelia with few septa across the hyphae and with asexual spores produced in sporangia are phycomycetes of the order Mucorales (see Fig. 19). Typical genera are:

1 **Mucor** — sporangia arise singly from the mycelium: especially common in soil.

2 **Rhizopus** — sporangia are produced at nodes of stolons: spores of this genus are common in the air and the organism is frequently responsible for rots on fruits.

3 **Absidia** — sporangia produced internodally along stolons: common in soil.

In cases where asexual spores are produced externally (conidia) and the hyphae of the mycelium are septate, representatives of the order Moniliales may be suspected (see Fig. 20). Typical genera are:

1 **Aspergillus** — slow growing blue-green colonies producing one-celled spores in chains radiating from the swollen head of the conidiophores.

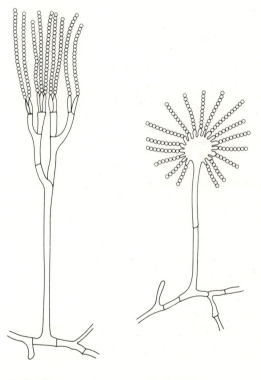

I *Penicillium* –
conidiophore with conidia

II *Aspergillus* –
conidiophore with conidia

Figure 20 Some common representatives of the order Moniliales (see also page 42)

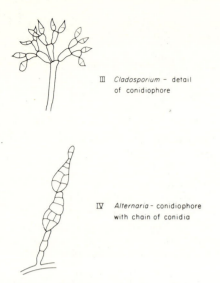

III   *Cladosporium* – detail of conidiophore

IV   *Alternaria* – conidiophore with chain of conidia

Figure 20 cont'd

2 Penicillium — similar in appearance to *Aspergillus* but easily distinguished as the one-celled spores are produced in chains from the branched ends of the conidiophores.

3 Cladosporium — slow growing colonies, often olive green in colour. Its spores are one or two celled, irregularly shaped and produced in chains like those of *Penicillium.*

4 Alternaria — a rapidly growing fungus with dark coloured club-shaped conidia borne in short chains. Each conidium is three to seven celled.

5 Candida — conidia budded directly from the sides of the hyphae giving a yeasty appearance.
(The genera *Cladosporium, Alternaria* and *Candida,* commonly develop on plates exposed to aerial contamination.)

6 Botrytis — fast growing grey mould with one-celled conidia borne in bunches on branched conidiophores. Common on decaying plant material. In culture hard black sclerotia are often produced in the medium.

7 Trichoderma — rapidly growing colonies, often smelling of coconut, with one-celled, bright green conidia borne in small slimy drops on branched conidiophores. Often common in soil where it may be

42

parasitic in other moulds like *Mucor.*
Some useful references on identification are 27, 31, 37 and 40.

GROWING FUNGI IN CULTURE

Cultures may be grown in media in Petri dishes or on slopes in tubes. Ordinary malt agar is useful for the isolation and culture of many fungi or alternatively an agar based on vegetable extracts can be used. Sporulation is usually encouraged if the cultures are exposed to twelve hours of light per day.

Preparation of agar

This is best done in bulk and for most purposes a 500 cm³ conical flask is a convenient preparation container. For malt agar thoroughly mix:

6.0 g malt extract

4.5 g agar

in 300 cm³ of water. Tap water can be used but where it may have stood in copper piping for any length of time it is usually better to use glass distilled water.

And for vegetable agar mix:

60 cm³ vegetable juice (a mixed vegetable concoction or V-8 vegetable juice manufactured by Campbell's Soups is suitable)

1.0 g calcium carbonate (to neutralise the acidity of the juice)

6.0 g agar

in 240 cm³ of water.

Agar will not dissolve in cold water and so the preparation container is stood in a boiling water bath until it has all gone into solution. Agar must never be added to boiling water (see page 9).

Preparation of slopes

McCartney bottles (8 x 3 cm with screw caps) or rimless test tubes 12.5 cm long, 1.5 cm dia., are convenient containers for fungal cultures. 5.0 cm³ of hot prepared agar should be placed in each test tube taking care to avoid getting any on the upper 3.0 cm of the tube. If the sides do become contaminated, they must be cleaned off before plugging the tube with a shaped, non-absorbent cotton wool cylinder. The cylinder should fit firmly occupying about 2.5 cm of the inside of the tube and projecting about 1.0 cm from the top. A properly made plug will, after sterilisation, prevent contamination of the inside of the tube as long as it is kept dry. A McCartney bottle

Safe Handling of
Micro-organisms

philip
harris

NOTES ON THE SAFE HANDLING OF MICRO-ORGANISMS

Although the micro-organisms we supply for use in schools are considered to be non-pathogenic to man, pathogenic mutants and pathogenic contaminants may appear. It is most important therefore that all cultures are treated as if they were pathogenic.

Infection may occur in the following ways:

A.   **INHALATION** of aerosols—water droplets containing bacteria. These are formed when culture vessels are opened carelessly and organisms transferred incorrectly.

   REMEDY—Use aseptic techniques (see Section II).

B.   **SKIN CONTACT**—as a result of breakages or spillages.

   REMEDY—Swab with antiseptic (see Section III).

C.   **INGESTION** as result of pipetting cultures by mouth.

   REMEDY—Never pipette orally always use teat pipettes.

I GENERAL SAFETY RULES

As well as following normal procedures for laboratory safety, the following points should be brought to the attention of the students.

1.   Wear a protective coat.
2.   Wash hands and cover any cuts with waterproof dressings.
3.   Avoid all hand to mouth operations (e.g. licking labels) and do not eat or drink in the laboratory.
4.   Follow the correct aseptic techniques when transferring organisms (see Section II).
5.   Provide a container of 2% Chloros for the disposal of used pipettes, culture vessels, etc. (see Section III).
6.   Do not isolate from potentially dangerous sources such as human mucus, cuts, etc. Similarly, do not subculture unknown organisms—they may be pathogens.
7.   Swab spillages immediately.
8.   Label cultures and seal for incubation with adhesive tape. If it is necessary to open the cultures after incubation, a filter paper dish soaked in 40% formalin should be placed in the lid of the disc 24 hrs. prior to examination.
9.   When the work is completed, dispose of all contaminated equipment as indicated in 5, swab the bench and wash hands before leaving the laboratory.
10.   While cultures can be kept in a refrigerator to increase their 'life', care should be taken that such laboratory refrigerators are not used to store food for human or animal consumption.

2

II ASEPTIC TECHNIQUES

Bacteria and fungi are generally grown on a special medium in either a screw topped universal container or McCartney bottle or in a petri dish. These containers are transparent and allow the growth of the colony to be

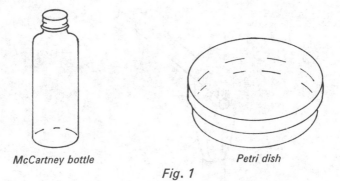

McCartney bottle   Petri dish

Fig. 1

observed. The micro-organisms are handled by means of an inoculating loop.

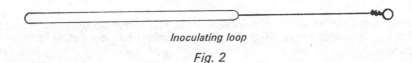

Inoculating loop

Fig. 2

Because bacterial and fungal spores are to be found in the air around us, on our clothes and bodies, in dust, etc., it is important when handling cultures of micro-organisms to follow set procedures. These procedures have been designed to minimise the risk of contamination both from and to the environment and are generally known as 'aseptic' techniques. These techniques should be practised several times as a 'dry-run' before working with cultures, so that any initial awkwardness can be overcome.

A bunsen burner or spirit burner is required for the sterilisation of the inoculating loop.

1. Select an area of bench away from draughts and open windows.
2. Swab the banch top with 70% alcohol or a working solution of Harris B.A.S. Cleaner.
3. Assemble the apparatus required.
4. Work near a bunsen or spirit burner as the up draught from the flame will help prevent contamination.

Note: A suitable transfer chamber may be used instead of the open bench. The Harris Transfer Chamber has been specially designed for school use.

3

Sterilisation of Inoculation Loop

This is done by heating the loop to **red heat** in a bunsen flame. It is important not to just heat the tip, but to heat it right back to the handle by moving it up and down in the flame.

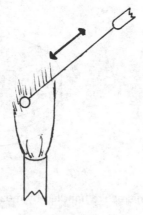

Fig. 3

After sterilisation, allow the loop to cool for a few seconds before use. After use, **always resterilise** the loop and allow to cool before putting it down.

Opening culture bottles

When culture bottles are open there is a risk of contamination from the atmosphere. The following procedure has been devised to minimise such contamination.

(i)   Loosen the cap, but do not remove it.

(ii) Hold bottle in left hand and remove cap with little finger of right hand.

(iii) Flame neck of bottle by passing neck through a bunsen or spirit burner flame, backwards and forwards. At this stage, something could be introduced or removed from the culture with an inoculating loop.

(iv) Flame neck again as in stage (iii).

(v)   Replace cap.

Remember to keep the time the cap is removed to a minimum.

Petri dishes

The plastic dishes obtained from suppliers have been sterilised by gamma irradiation. The pack should be kept intact as long as possible. On opening the pack, remove only the number of petri dishes you require and reseal the pack by folding over and securing with adhesive tape.

4

The inside of the dish will remain sterile as long as the top remains in position. As there are a few spare dishes in the pack, it is worthwhile removing one to practice the technique below.

It will be necessary to pour medium into dishes and to introduce and remove organisms using the inoculating loop. When this is being done, the lid of the petri dish should be lifted off with the left hand and held at an angle over the dish. The operation can then be performed with the right hand. Do not completely remove the lid, and replace it immediately the operation is complete.

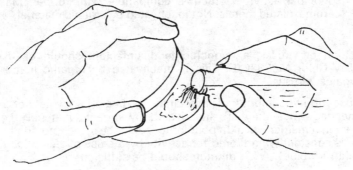

Pouring a plate

Fig. 4

Practice:

(i) flaming the loop, lifting the petri dish lid, placing the loop inside, removing it, replacing the lid and reflaming the loop.

(ii) holding empty bottle in right hand, removing cap with left hand and placing on bench, flaming neck, lifting petri dish lid, pouring from bottle, replacing petri dish lid and replacing cap.

III DISPOSAL

It is most important that no living culture or contaminated equipment is disposed of without first being made harmless by sterilisation. Procedures are given below.

1. Sterilisation by autoclaving

This method relies on steam at 121°C to kill off the micro-organisms. if you do not possess an autoclave, then a domestic pressure cooker will make a suitable substitute. Glassware may be autoclaved directly, but first loosen caps on screw-topped containers. Plastic materials should be placed in the disposal bags provided and then sealed. Autoclave the items at 15 lb/sq. in. pressure for twenty minutes, after which glassware may be cleaned, washed and re-used. The disposal bag should, ideally, be incinerated, but may otherwise be discarded with laboratory rubbish.

2. Sterilisation by soaking or steeping

In this method, chemicals are used to kill the organisms :

(a) Harris Lab Disinfectant M86810/9.
    A clear phenolic disinfectant recommended in the Howie Report for bacteriological work. It is not inactivated by organic material. Not suitable for viruses.

(b) Harris Cold Sterilising Fluid M86870/6.
    A broad spectrum disinfectant with bactericidal, virucidal and sporicidal activity, which is compatible with rubber, plastic and chrome-plated metals. Not inactivated by organic material.

Note

For blood or viruses, a hypochlorite distinfectant should be used, e.g. Harris Dry-Chlor M86830/5. This is inactivated by organic matter, but is active against viruses.

The use of hypochlorite disinfectants for bacteriology is not normally recommended since they are inactivated by organic material (i.e. agar). However, in emergency, a hypochlorite could be used provided that the solution is constantly monitored for activity by the use of starch iodide paper (it should turn blue). A 2% dilution should be sufficient.

SPILLAGE

Laboratory spills should be soaked up as they occur with Harris Lab Spills powder (M86850/0).

Microbiological equipment and cultures

An extensive range of microbiological equipment and cultures is available from Philip Harris Biological and details can be found in our current catalogue. In addition, the following visual aids and books are recommended.

Filmstrips

A10685/9   Safe Handling of Micro-organisms.
A10687/2   Dispersal of Micro-organisms.
A10689/6   Viruses.
A10691/4   Penicillin.
A10693/8   Methane Production.
A10697/5   The Mushroom.

Books

A93700/9   Micro-organisms by Williams and Shaw (Mills & Boon 1976).
A94250/9   Bacteriology by J. Humphries (John Murray 1974).
A93013/1   Micro-organisms by P. F. Fry (Hodder & Stoughton).

Imperial College Safety Booklet: Precautions Against Biological Hazards 1974, available from Imperial College, London SW7 2AZ.

Useful addresses

Professional advice on the use and handling of micro-organisms is available from:

The Institute of Biology (MISAC), 41 Queen's Gate, London SW7 5HU.

The Association for Science Education, College Lane, Hatfield, Herts. AL10 9AA.

The Scottish Schools Science Equipment Research Centre, 103 Broughton Street, Edinburgh EH1 3RZ.

Printed by Burman, Cooper and Company Birmingham B8 2SG

Philip Harris Biological
Oldmixon, Weston-super-Mare, Avon BS24 9BJ.
Telephone (0934) 413063, Telex 449248
A Division of Philip Harris Ltd.

requires 10 cm³ of agar mixture and should be sterilised and cooled with the cap loosened.

After sterilisation (103.5kN/m² for 15 minutes) the tubes or bottles should be allowed to cool so that the agar solidifies as a slope (Fig. 3). Slopes in tubes can be kept for some time but the agar does dry out quickly unless they are stored in a sealed container like a polythene bag. Unfortunately, the plugs may then become damp and cease to act as efficient filters. Screwed down McCartney bottles will keep indefinitely but before use the cap and adjacent glass should be swabbed with alcohol and lightly flamed.

Preparation of plates
A standard Petri dish requires 15 cm³ of medium and should be prepared in exactly the same way as the procedure used to make plates for the growth of bacterial colonies. (See Chapter 1, page 10.)

Inoculation of slopes or plates from agar cultures
Hold the agar culture and a slope in one hand and an inoculating needle in the other. Sterilise the needle by heating to redness in a flame, allow it to cool while removing the plugs or caps from the tubes or bottles; pass necks of both through a non luminous flame. Cut a small piece of agar (5 mm² is ample, but it can be much smaller) carrying the fungus from the culture, and transfer it as quickly as possible to the fresh agar surface, flame the necks of the tubes or bottles again and replace the plugs or caps. Incubate at about 20—25°C. Within a day or so the fungus should grow out from the inoculum and then spread over the fresh surface.

To inoculate an agar plate, place it on a clean and, if possible, slightly damp surface. Then using the same procedure as for a slope, transfer a piece of agar with fungus to the centre of the agar plate. Expose the new agar surface to the atmosphere as little as possible during transfer. After incubation, the fungus will start to form a circular colony.

HANDLING FUNGAL CULTURES, ESPECIALLY DRY SPORED SPECIES
When handling cultures it is important to remember that most fungi produce spores which are dispersed aerially and that these spores may remain viable for many weeks.[13, 23] Obviously pathogenic species like *Aspergillus fumigatus* (a cause of pulmonary disease) should not be used in teaching.

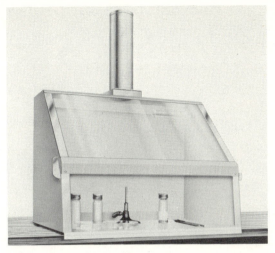

Figure 21 A transfer chamber (Photo by courtesy of Philip Harris Biological Limited.)

Furthermore in order to reduce the incidence of contamination in a laboratory, it is important that care should be taken when examining cultures, particularly of fast-growing forms (like species of *Penicillium*, *Aspergillus* and *Rhizopus*).

To avoid contamination take the following precautions and see page 13:

1 never open plates unnecessarily. It is best to keep them sealed with self adhesive tape.
2 when it is essential to open them, do this *slowly* without jerking.
3 do not pour or open plates in draughts. Air movements disperse spores. Plates should only be opened within an inoculating or transfer cabinet (Fig. 21).
4 when collecting the spores in a loop, always use a wetting agent and again avoid jerky movements.
5 inoculate plates with the surface of the medium facing downwards to reduce the chances of contaminant spores settling on the surface.
6 touch spores onto the surface and avoid 'flicking' any across the plate.

A useful method of reducing the amount of aerial contamination is to spray the atmosphere (using a small hand sprayer) with a disinfectant solution or a 70% solution of methylated spirits. Particular care is needed if methylated spirit is used since such a solution can ignite and for school work the disinfectant solution is best.

METHODS OF EXAMINING CULTURES

A great deal of useful information may be obtained from microscopic observations of cultures in situ. Examination in this way should always precede the preparation of slides. Petri dishes may be placed on the microscope stage either side up and examined with the lowest power objective so that the mould may be viewed directly from above or beneath through the medium. Vital information for identification, whether stolons are present for example, can only be obtained by studying undisturbed cultures.

When examining cultures it is important to remember that a good deal of water is exuded from the aerial parts of certain moulds and that some of this may appear as droplets. This is common on the sporangiophores of species of *Mucor* and *Phycomyces.*

1 Slide cultures

The fruiting bodies of some moulds are too small to be able to discern detailed structure with lower power objectives and too fragile to prepare successful slides. In these cases it is often useful to make slide cultures of the fungus. These are prepared by inoculating the edges of a cylinder of medium on a slide in a humidity chamber constructed from a Petri dish as shown in Figure 22.

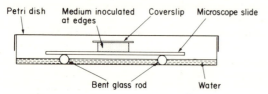

Figure 22 Slide culture using a Petri dish as a humidity chamber

2 Culture on media with low concentrations of nutrients

A convenient way to study the mycelium and asexual structures of many fungi is to grow them on media containing a low concentration of nutrients, for example, media containing 0.2–0.4% malt extract and 1.5–2.0% agar. A sparse growth of the mould results and often the manner of spore formation may be easily observed. Small squares of the culture may

be cut out, mounted on a slide, stained and examined, or the whole culture may be examined in situ. Drops of immersion oil can be placed directly upon the culture, which may then be observed using an oil immersion lens. This is a particularly good way of observing protoplasmic streaming. Growth of fungi on such media is always sparse and may sometimes be abnormal and so it is always advisable to have plates of the mould grown upon normal media available for comparison.

PREPARATION OF SLIDES

For study of fine details and accurate measurements of fungal structures, slide preparations are needed.[4],[6] It is a general practice to use fluid mounts made with as little manipulation as possible.

The mounting fluids used include:

1 alcohol, which wets fungal structures effectively and is a satisfactory mountant for brief examinations.
2 lactophenol, which is probably the most useful general mountant (composition, phenol 10 g, lactic acid 10 g, glycerol 20 g, distilled water 10 g).
3 glycerol.
4 water, which is better than lactophenol for mounting yeasts.

1 Method of mounting

Place a *small* drop of mounting fluid in the centre of the slide. Pick off a small portion of *typical* material from the culture, place this in the drop of mountant and very gently tease it out with a pair of needles until it is well wetted, lower a cover glass on to it in such a way as to avoid air bubbles. Mop up any excess mounting fluid with bits of filter paper.

2 Staining

It is often useful to stain the material before mounting it. A number of stains have been used, including the following:

1 cotton-blue in lactophenol, which is the most useful stain (0.05 g cotton-blue in 100 cm^3 lactophenol).
2 trypan blue in lactophenol (0.25 g trypan blue in 100 cm^3 lactophenol).

The specimen may be mounted directly in the stain, but better results are obtained if after staining for about half a minute the material is then mounted in plain lactophenol. Excess stain may be drawn off with bits of filter paper, or alternatively the material may be stained on one slide (or in a watch glass) and then transferred to a clean slide to be teased and mounted.

3 Permanent slides

Any slide which is to be kept as a permanent record should be protected by sealing the edge of the cover glass with a suitable cement. Cements which can be used to make slides of moulds mounted in lactophenol permanent include brown shellac cement and ordinary nail varnish. The cement should be of a consistency that runs smoothly but not too readily from a brush. A very thin coat of cement should be put on first, and when this is dry a second thicker coat is added to overlap the first.

SOME INVESTIGATIONS WITH PLATE CULTURES

The scope for investigations of fungal characters using plate cultures is literally without limit but the following list gives some examples which can be educationally valuable.
1 Assessment of growth by measuring increase in colony diameter, or by measuring area in various ways. This needs to be done daily, or even more frequently with fast growing fungi.
2 Growth at different temperatures. This enables the skewness of the growth—temperature curve to be readily appreciated.
Comparison of growth rates of different fungi; for example, fungi imperfecti compared with phycomycetes like *Mucor*.
3 Induction of mutants.

1 Growth of fungi on solid media

Measurement of linear growth is a popular and easy way of measuring fungal growth. It is best to use colony radial growth rate in preference to colony diameter as this has been found to be a reliable parameter to use in comparing growth of a mould both at different temperatures or on media containing different concentrations of a growth inhibitor. However, it cannot be reliably used to determine the effect of nutrient concentration on growth or to compare the growth of different species or even mutants of the same species.

To carry out an investigation prepare spore suspensions of suitable fungi, like *Neurospora crassa*

or *Cunninghamella*, by washing seven day slopes of the fungus with a little sterile distilled water. Inoculate dry malt extract test plates centrally with spore suspension dispensed from a sterile Pasteur pipette. It is advisable to load three separate plates for each species in order to average the data. Incubate all the plates at $30°C$ and measure the diameter of each colony (by two readings at right angles) at 0, 2, 4, 6, 24, 48, 72 hours after inoculation. After calculating the mean colony diameter at each period a graph of colony diameter against time can be used to summarise the data.

2 Determination of the optimum temperature for fungal growth

Again prepare spore suspensions from seven day old slopes and inoculate malt extract plates centrally. A useful temperature range is $20°$, $25°$, $30°$, $37°$ and $45°C$ achieved in incubators with a refrigerator at $4°C$ for comparison. It is best to incubate three plates at each temperature. After eighteen and thirty-six hours of incubation measure the diameter of each colony (taking two readings at right angles) with a plastic ruler and a dissecting lens. Calculate the mean colony radial growth rate at each temperature per hour and plot a graph of temperature against growth rate.

Suitable fungi to use are *Rhizopus* and *Cunninghamella*.

3 Mutation in Aspergillus nidulans [AL]
(a) Mutagenesis

As a result of natural events spontaneous mutation occurs in all organisms but it is technically difficult to detect in most cases. A variety of treatments can be used to increase this natural background level of mutation. One of these is irradiation with ultraviolet light which is strongly absorbed by deoxyribonucleic acid. This effect may be demonstrated by the production of mutant colonies of *Aspergillus nidulans* following irradiation of uninucleate spores. It is possible to detect both spore colour mutations and morphological mutants.

To carry out the demonstration the following equipment is needed.
An ultraviolet lamp emitting 80% of ultraviolet output at 253.7nm (2537 Å)
500 cm^3 sterile Ringer's solution—see page 27 (or 0.85% sodium chloride or even distilled water)
50 cm^3 0.1% detergent

45

40 plates of malt agar

Aspergillus complete medium or any medium suitable for fungal growth like malt agar.

A suitable sporulating agar for *Aspergillus* can be prepared from:

Sodium nitrate (NaNO$_3$)	0.1 g
Magnesium sulphate (MgSO$_4$.7H$_2$O)	0.05 g
Potassium chloride (KC1)	0.05 g
Potassium dihydrogen orthophosphate (KH$_2$PO$_4$)	0.15 g
Glucose	2.0 g
Agar	1.5 g

100 cm^3 of distilled water

—with the addition of the slightest trace possible of iron (II) sulphate and zinc sulphate.

(0.1% sodium deoxycholate may be added to restrict colony spread)

Haemocytometer slide

Glass spreader and alcohol in Petri dish

50 sterile plugged test tubes

5 10 cm^3 graduated pipettes

10 1 cm^3 graduated pipettes

5 slopes of *Aspergillus* with green spores

Mounted loops

A suitable procedure is:

1 set up the u.v. lamp at a height of 30 cm above the working surface to be used and leave it switched on for 30 minutes before starting the experiment. As always, *do not look directly at the u.v. light as this can cause damage to the eyes.*[9]

2 make a 10 cm^3 suspension of conidia in diluted detergent—about 0.1% is convenient and a concentration of about 10 spores per cm^3 of liquid is ideal. (Virtually any detergent can be used but some domestic varieties do kill fungi and bacteria. A trial and error process can be used to discover washing-up liquids which work or the proprietary brands 'Tween 80' or 'Triton X 100' purchased from laboratory suppliers.) This is done by taking a loop wetted with detergent and streaking it over the surface of a slope about three times. The suspension should be a visible pale green colour and if it is not repeat the process until it is.

3 agitate the suspension to separate conidia by drawing it up and down into a pipette.

4 estimate the number of conidia per cm^3 by means of the haemocytometer slide.

5 prepare two sets of five tubes each containing 9 cm^3 of Ringer's solution and label one set (which will be used for irradiated spores)

$I^{1\,0-1}$ $I^{1\,0-2}$ $I^{1\,0-3}$ $I^{1\,0-4}$ $I^{1\,0-5}$

and the other set (which will act as a control series)

$C^{1\,0-1}$ $C^{1\,0-2}$ $C^{1\,0-3}$ $C^{1\,0-4}$ $C^{1\,0-5}$

6 serially dilute the untreated spores by placing 1 cm^3 of suspension into the 9 cm^3 of Ringer's solution in the tube marked $C^{1\,0-1}$, then take 1 cm^3 from the tube and place it in the tube marked $C^{1\,0-2}$, mix and continue the process down to $C^{1\,0-5}$.

7 spread 0.1 cm^3 of $C^{1\,0-5}$ $C^{1\,0-4}$ and $C^{1\,0-3}$ on three of the agar plates.

8 place the remainder of the detergent spore suspension in a sterile Petri dish. Remove the lid and place it on a sheet of white paper under the u.v. lamp and expose it for one minute.

9 dilute the treated suspension 10^{-1}, 10^{-2}, 10^{-3}, 10^{-4}, 10^{-5} and place 0.1 cm^3 of each dilution on 5 agar plates.

10 repeat the whole process with exposure times of 2, 4, 10 and 20. (In the classroom situation a class of students can be conveniently organised as five groups, each group making up one set of materials for each of the exposure times.)

The dilution part of the experiment may be simplified by streaking the control and irradiated suspensions on two separate plates. Take a loopful of appropriate spore suspension and streak it at one side of the plate. Flame the loop and cool it in sterile distilled water. Then, carry out further streaks II–IV (see Fig. 6 page 13).

Note that it is best for the streaks to be further apart than in bacterial streaks. This method will be non-quantitative but should show the production of a few mutants.

Incubate plates for 3 days at 37°C or 6 days at room temperature and then examine for differences in spore colour and morphological form.

Morphological mutants usually occur in the approximate proportions shown in Table 10.

The viability of fresh spores should be 100%.

u.v. dose in mins	survival %	% mutants
1	70	1
2	20	9
4	15	15
10	1	20
20	0.1	10

Table 10 Approximate proportions of morphological mutants

(b) Nutritional Mutants

The fungus *Aspergillus nidulans* can grow on a very simple medium containing nitrate as nitrogen source, glucose as carbon and energy source, and a variety of mineral salts. This is the so-called minimal medium (MM) prepared from:

Sodium nitrate ($NaNO_3$)	6 g
Potassium chloride (KCl)	0.52 g
Magnesium sulphate ($MgSO_4.7H_2O$)	0.52 g
Potassium dihydrogen orthophosphate (KH_2PO_4)	1.52 g
Iron II sulphate ($FeSO_4.7H_2O$)	small crystal about 1 mm^3
Zinc sulphate ($ZnSO_4.7H_2O$)	small crystal about 1 mm^3
Glucose	10 g
Agar	15 g
Distilled water	1 litre

After mixing adjust the pH to 6.5 with hydrochloric acid and sodium hydroxide. Autoclave at 69kN/m^2 for ten minutes. The light precipitate which will form after autoclaving does not affect fungal growth. It is important to avoid over autoclaving as excessive precipitation and contamination will occur.

Following mutagenic treatment with ultraviolet light or chemicals, strains can be isolated which fail to grow on minimal medium. Such strains can grow, however, when the minimal medium is supplemented with, for example, vitamins or amino acids.

Such strains are nutritional mutants in which single enzymes in biosynthetic pathways are defective. This results in the inability of the strain to synthesise essential growth requirements. These compounds must now be supplied in the medium. Normally, each mutant strain has a single

requirement, supporting the one gene-one enzyme hypothesis. Strains with more than one requirement can be selected from the progeny of crosses between strains which require different growth factors.

The behaviour of such strains may be demonstrated with:

Slope cultures of *Aspergillus* strains requiring:
1 p-amino benzoic acid (paba).
2 adenine (ad).
3 adenine and p-amino benzoic acid (ad, paba).
4 wild type strain—no requirements (ad^+ $paba^+$).

Take four plates each of minimal medium with:
1 no additions (MM).
2 150 $\mu g/cm^3$ adenine (MM + ad).
3 200 $\mu g/cm^3$ p-amino benzoic acid (MM + paba).
4 100 $\mu g/cm^3$ adenine and 200 $\mu g/cm^3$ p-amino benzoic acid (MM + ad + paba).

A tube containing 2 cm^3 0.1% detergent to act as a wetting agent. Mounted loops.

A convenient practical procedure is:
1 arrange the four plates so that the medium surface faces downwards.
2 label segments of each of the plates at appropriate places with the four strain types (see Fig. 23).

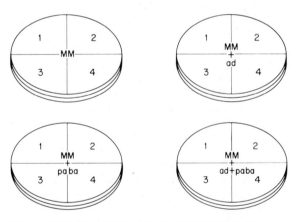

Figure 23 Culture plates labelled MM, MM + ad, etc.

3 flame a nichrome loop and cool in the tube of 0.1% detergent. Remove a loopful of the dilute detergent and use this to pick up spores from the first slope.
4 inoculate the first strain at the appropriate place on each of the four plates.
5 repeat the procedure with the three further strains.

6 incubate the plates, again with the medium surface facing downwards, at 37°C and examine growth at 24 and 48 hours (alternatively at 25°C for 2—4 days).

The results may be scored on a grid of the type shown in Table 11.

		MM	MM + ad	MM + paba	MM + ad + paba
paba	1	—	—	+	+
ad	2				
ad paba	3				
ad^+ $paba^+$	4				

Key: — no growth
+ growth

Table 11 A grid to score the growth of nutritional mutants of Aspergillus on differing media
(The results for a p-amino benzoic acid requiring stain are indicated.)

GROWTH OF FUNGI IN LIQUID CULTURE

Because agar is a natural product of somewhat uncertain composition even after purification, cultures on agar are not used in critical physiological and biochemical experiments with fungi. Instead, liquid media, of defined compositions, are used. The technique may be conveniently demonstrated using the common soil saprophyte *Aspergillus niger,* which competes well with contaminants. Further information may be found in references 6, 10 and 33.

1 An investigation of the growth of Aspergillus in liquid cultures of differing compositions

Prepare the following four nutrient solutions by dissolving the component(s) for each in $250cm^3$ distilled water:

Carbon Source (C) — Glucose 20 g
Nitrogen Source (N) — Sodium nitrate
$(NaNO_3)$ 1.0 g
Phosphorus Source (P) — Potassium dihydrogen orthophosphate
(KH_2PO_4) 1.0 g
Salt Source (S) — Potassium chloride (KCl) 0.25 g
Magnesium sulphate
$(MgSO_4.7H_2O)$ 0.25 g

Iron (II) sulphate
$(FeSO_4.7H_2O)$ 0.001 g

Then measure into standard medicine bottles (284 cm^3 or 10 oz size) the volumes of each solution as indicated in Table 12.

Solutions Medium	C	N	P	S	Water
1 Minus carbon	0	8	8	8	8
2 Minus nitrogen	8	0	8	8	8
3 Minus phosphorus	8	8	0	8	8
4 Minus salts	8	8	8	0	8
5 Complete medium	8	8	8	8	0

Table 12 Deficient media for liquid culture of Aspergillus. Volumes in cm^3

In order to obtain reliable data it is best to make up each medium in triplicate.

Cap each bottle loosely or plug the bottles with coloured non-absorbent cotton wool and label them with a tie-on pencilled label. It is undesirable to use chinagraph type pencil markings as they are removed by sterilisation and many of the more permanent markers should be avoided as these are difficult to erase after sterilising. Sterilise at approximately $69kN/m^2$ for 10 minutes, and allow to cool.

Prepare a spore suspension by adding about 10 cm^3 of wetting agent solution to a slope culture and suspend the spores by rolling the culture vigorously between the hands while keeping it vertical. Store the suspension in a 28 cm^3 (1 oz) bottle. After allowing the medicine bottles to cool, inoculate each bottle of medium with one drop (about 0.1 cm^3) of suspension from a sterile Pasteur pipette. All this should be done aseptically in a draught free room.

Incubate the culture bottles flat on their largest sides at room temperature or in an incubator at between 25°C—37°C. The temperature is not important but can be chosen to suit the period before the mycelium is harvested.

At 37°C incubate 3 (or 4) days
30°C   5—6 days
25°C   7—8 days
room temperature, 7—10 days.

If incubation is too brief the differences in weights will be small but if harvesting is delayed and autolysis has occurred there will be a substantial reduction in weight. This reduction may well be most marked in the cultures with the highest growth rates.

To harvest the cultures, first add a few cm^3 of 0.1—1.0% detergent solution and shake gently to wet the spores and prevent contamination of the laboratory. Take a weighed cup made by shaping aluminium foil around the base of a 2.5 cm (1 in) specimen tube. Pierce a few holes in the bottom of the cup with a fine needle and collect the mycelium from each bottle separately. Dry each to constant weight at between 65°—105°C and determine the weights of the mycelia. The dried mycelia are hygroscopic, so they should be kept in a desiccator whenever possible.

2 Other growth investigations with Aspergillus in liquid culture

Further investigations which raise interesting problems are:

1 measure growth by harvesting cultures on complete medium after different times. This will illustrate possible dangers of assessing growth by taking dry weights after only one period of incubation.

2 compare growth at different temperatures.

3 measure growth on complete medium at normal strength and with some other concentrations in the range x 8, x 4, x 2, x 0.5, x 0.25, x 0.125, of each of the constituents.

4 compare growth with different sources of nitrogen. The sodium nitrate in the complete medium may be replaced by—

Ammonium nitrate (NH_4NO_3)	0.55 g
Ammonium chloride (NH_4Cl)	0.75 g
Ammonium succinate	1.05 g
Glutamic acid	2.0 g

Metabolism of these different compounds can lead to pH changes in the culture so the different media should first be adjusted to pH 5.5—6.0 before sterilisation so that there is a known starting value. In addition, if *A. niger* is used some of the characteristic effects may be masked by the fact that this fungus readily produces acid in culture. It is also instructive to use members of the Mucorales for investigations of this sort since they cannot use nitrate but do use ammonium ions or compounds with an NH_2 group attached in the molecule.

Some useful background sources of information are references 42 and 48.

THE KINETICS OF SOME BIOCHEMICAL PROCESSES TAKING PLACE AROUND YEAST CELLS

The dough fermentation processes in which *Saccharomyces cerevisiae* causes a dough to rise as a result of the release of carbon dioxide, is one which may be conveniently used to demonstrate the process of a biochemical process over a period of time.[29,35] Fortunately, the induction or lag period often necessary for a yeast inoculated dough to evolve carbon dioxide may be shortened by adding glucose to the flour and by increasing the amount of water usual for dough making so that a more fluid mixture is produced. Because of the protein in the flour (gluten), the surface of this mixture is relatively tough and will retain carbon dioxide during the early stage of leavening. However towards the end of the fermentation process, the surface may break due to the proteolytic enzymes in the yeast degrading the gluten and so carbon dioxide escapes.

A class investigation which may be completed in one hour can be done by providing groups of students with:

Glucose	5 g
Plain flour	100 g
Baker's yeast (cake)	7 g
Measuring cylinder	1 litre
Graph paper	

They are then instructed to:

mix together the glucose and plain flour. Disperse the yeast in 120 cm^3 of water at 45°C and stir this suspension into the flour to form a smooth paste. Pour the paste into the measuring cylinder and, as soon as the paste surface is level, note its volume. Take this as the zero time reading. Then record the volume of the paste at 3—5 minute intervals until the volume remains constant and finally, plot a graph of volume change against time. If high protein (pastry type) flour is not available the surface of the paste will soon rupture. This may be prevented to some extent by the addition of about 2 g of sodium chloride to the 100 g of flour or by increasing the amount of flour. Should fresh yeast be unavailable, dried yeast can be used. It should be prepared for each investigation by stirring 6 g of dried yeast and

49

8 g of glucose into water. The mixture should stand in a warm place at 45°C for about 15 minutes before adding the flour to make the experimental paste.

This simple investigation is capable of interpretation at many levels. It provides a convincing demonstration of a change in a system with time due to a chemical reaction and develops the graphical treatment of observations. The fact that the catalyst in the reaction is an enzyme can be proved by running a parallel experiment incorporating heat killed yeast and a cell free extract of the yeast can be used as well. This may be prepared by taking a thick suspension of yeast in water and grinding it in a chilled mortar with acid washed sand. The slurry is then filtered clear and the filtrate mixed with the glucose—flour mix.

At a more advanced level the kinetics of the catalytic system can be explored by varying such parameters as the substrate concentration (glucose), the enzyme concentration (yeast) and the temperature. Fortunately during the short period of the investigation, little growth of the yeast is to be expected therefore complications due to a changing amount of catalyst are eliminated.

The complete fermentation of 5 g glucose yields approximately 1400 cm^3 of carbon dioxide at 37°C. Thus, from the observed increase in volume of the paste an estimate of the percentage conversion of the glucose can be obtained. At the 36% conversion level the ethanol concentration is 1.5% and at this level it begins to exert an inhibitory effect on the reaction. This is one of the explanations of the falling rate of volume increase in the final stage of the investigation.

FUNGAL PATHOGENS OF PLANTS [AL]

Students can be led to appreciate the full significance of the ravages wrought by fungal parasites in a variety of ways. In the teaching laboratory two instructive approaches are to present students with diseased plant tissue asking them to find the causal agent and to explore the conditions under which it may cause infection, given a pure culture.

Principles and a Method for Investigating the Causal Agents Responsible for Fungal Rots

In 1890 Robert Koch laid down exacting principles to be used to decide whether an organism

50

can be regarded as the cause of a specific disease. Briefly his postulates state:

1 the organism must be found in every case of the disease
2 it must not be found as an accidental organism in other diseases
3 after it has been isolated into pure culture from the diseased organism it must have the ability to produce the identical disease in organisms of the same species.

A procedure which may be applied to some infected material to establish these principles is:

1 examine an infected plant tissue and record the disease symptoms.
2 with a sterile scalpel cut away the most badly damaged tissue, which may be contaminated by secondary invaders, and try to expose the advancing front of the disease. Take a little of the tissue from behind the advancing edge of the lesion and macerate it in sterile distilled water in a tube. Leave for about 10 minutes.
3 shake the tube and streak out heavily on nutrient agar and malt extract agar plates and incubate at room temperature.
4 subculture colonies from the above plates by streaking out on to fresh agar plates.
5 compare the morphology of your isolates with that of the organism in the original infected plant tissue.
6 wash some uninfected plant tissue and swab it with methylated spirit. Allow the alcohol to evaporate and wound the surface with a sterile scalpel and inoculate with the isolated organism. Store treated material in sterile beakers as shown in Figure 24.

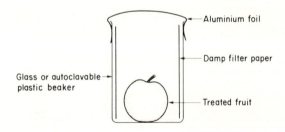

Figure 24 Beaker used to store inoculated plant material

7 determine if the disease symptoms of the inoculated plant tissue are identical with the original infected plant.

Some questions which may be used to establish infectious conditions

Students should be advised to think along these lines:

1 is the presence of a wound on the surface of the plant a pre-requisite for infection. Most soft rotting pathogens only infect via a wound of some sort. Can it infect via a bruise?
2 what is the influence of relative humidity on the course of infection?

The following saturated salt solutions give the stated relative humidities at $20^{\circ}C$: magnesium chloride, 32%; sodium bromide, 58%; ammonium chloride, 77%; potassium chloride, 85%; sodium carbonate, 92%. Further information on this may be found in reference 10.

3 what is the influence of temperature on the course of infection?
4 what is the influence of inoculum size on infection?
5 what is the effect of competition between two pathogens on the same plant tissue?
6 what is the best range of conditions for the pathogen?

Host pathogen combinations suitable for class investigation

Some relatively easily obtained combinations are shown in Table 13.

Pathogen	Host tissue
Penicillium italicum	Lemon
P. expansum	Apples
P. digitatum	Orange
Pythium ultimum	Potatoes (watery wound rot)
Sclerotinia fructigena	Apple

Table 13 Some host/fungal pathogen combinations[8]

When asking students to carry out investigations with these materials it is essential to get them to plan exactly how they propose setting about their work and then have a general discussion before allowing them to proceed. The following important points are often neglected:

1 the need for controls. (For example, inoculating pathogens on to undamaged plant tissue as well as on to wounded areas of the plant and the need for a control of wounded but uninoculated plant!)
2 the use of aseptic techniques like washing the outside of the plant tissue with water and then swabbing it with cotton wool soaked in methylated spirit, cutting the surface of the plant with a *sterile* scalpel and storing the plant tissue during the investigation in an autoclaved beaker capped with aluminium foil.

INVESTIGATING DAMPING-OFF FUNGI [AL]

Over watered seedlings are very prone to fungal infection at the base of the stem so that rotting occurs at this site and the leaf shoots fall over. Contrary to popular belief this damping-off condition can be caused by more than one species and the inoculation of seedlings with damping-off fungi can give interesting teaching data.[28]

1 Raising test seedlings

Suitable seedlings for investigation can be raised from lettuce and cabbage seeds. They are best grown in 7–10 cm diameter plastic pots which have been filled to within 2.5 cm from the top with fine grade Vermiculite. This should be well soaked with half-strength proprietary liquid feed solution and about 100 seeds sown evenly in each pot. Finally, the seed should be covered with not more than 0.25 cm of moist sand and the tops of the pots sealed with thin transparent polythene film to prevent water loss. The pots should be kept in a warm place so that rapid germination takes place.

2 Inoculating seedlings

Take cultures of *Pythium debaryanum*, *Rhizoctomia solani* (crucifer strain) and *Corticium praticola* on a vegetable juice agar and when the lettuce and cabbage seedlings are about 1.0 cm high, place a substantial piece of agar with mycelium on the surface of the sand near the wall of each pot. Place some supports in each pot to raise the level of the covering polythene film at least 3.0 cm above the seedling, replace the film and set the pots aside for a few days. The disease will spread from the inoculum to all the seedlings but it is essential to remember that the initial agar piece must not dry out and that seedlings become resistant to these parasites as they get older. So for best results use young seedlings soon after emergence.

Typical damping-off symptoms are to be seen as in Table 14.

Fungus	Lettuce	Cabbage
P. debaryanum	+	+
R. solani (crucifer)	–	+
C. praticola (closely related to R. solani)	+	+

Table 14 Damping-off in lettuce and cabbage. A plus (+) indicates that the plants collapse and a minus (−) no change.

3 Isolation of fungi from damped-off seedlings

The isolation of parasites from damped-off seedlings presents certain difficulties because of the delicate nature of the infected tissues, and their liability to contamination by bacteria. Moreover the pathogen is very close to the surface of the host and surface sterilisation may well kill it.

A good method to use is to remove an infected seedling from the soil, cut off its roots and leaves, and wash the hypocotyl and stem in running tap water. Then remove 4–5 mm of the hypocotyl or stem, containing about equal parts of healthy and diseased tissue and transfer it with sterile forceps to 10 cm^3 of sterile water in a sterile Petri dish, and shake well; repeat this 4–5 times taking care to remove as much surplus water as possible on each transfer. Then place the tissue on dry, sterile filter paper for a short time to remove most of the surface water and finally, transfer the piece of tissue to the surface of a plain 1.5% agar in a plate and incubate at 20–25°C.

When colonies have grown, transfer hyphal tips to fresh plates of acid agar and incubate again. To prepare acid agar take:

Agar	1.5 g
Malic or tartaric acid	0.5 g
Distilled water	100 cm^3

Put the agar in a 250 cm^3 flask and add 90 cm^3 of water. Add the acid to 10 cm^3 water in a test tube. Plug the flask and the test tube and sterilise by autoclaving at approximately 103.5 kN/m^2 for 20

52

minutes. Cool both solutions to 47–50°C, add together and mix well. Pour into five or six Petri dishes.

Examine the new colonies and if these seem to be free from bacteria, transfer pieces from the edge to plates of potato-dextrose agar prepared from:

Cooked Potato	20 g
Agar	2.0 g
Dextrose	2.0 g
Water	100 cm^3

Any stray contaminating bacteria will grow on this medium and be revealed, though for a quick check, pieces of agar with hyphae can be transferred to tubes of nutrient broth and incubated overnight. When the cultures are free from bacteria, transfer small pieces from the edge of the colony to slopes of vegetable juice or potato-carrot agar. The latter may be prepared from:

Grated raw potato	1.5 g
Grated raw carrot	1.5 g
Agar	2.0 g
Water	100 cm^3

4 Methods for eliminating bacterial contamination

If, after treating a damping-off fungus as above, the colonies obtained are still contaminated with bacteria, then some other methods which can be used to get pure culture of the fungi are:

1 transfer the fungus to plain agar containing 30 p.p.m. Rose Bengal and 30 p.p.m. streptomycin. After incubation, subculture from the edge of the colonies produced.

2 transfer the fungus to acid agar. Warm a glass ring (10–15 mm in diameter) and put it round the inoculum so that it melts and sinks into the agar. The fungus will now grow through the agar underneath the ring and come to the surface some distance from the ring. Because most of the contaminating bacteria will be aerobic, there is a reasonable chance that hyphae on the surface outside the ring will be free from bacteria. Transfer a piece from the edge of the colony outside the ring on to potato-dextrose.

3 establish a colony on acid agar. Shake the agar so that it comes to rest on the lid of the Petri dish. Transfer a piece of agar from the new top surface (the original bottom) at the edge of the colony to a fresh plate of agar. In growing through the agar there is again a reasonable chance that the hyphae

will become free of bacteria by the time they have reached the bottom of the agar. Note that it is possible that the fungus causing the disease will not grow well, if at all, on plain agar or acid agar. If this is so, then some vegetable agar should be tried instead.

ROTTING FUNGI

A number of fungi produce cellulases which can break down cellulose. A common one which occurs in soil is *Chaetomium globosum* and its black perithecia are frequently to be seen on damp wallpaper and on tents put away before they have dried out thoroughly.[26] The fungus can be conveniently obtained by burying strips of cotton in the soil and subsequently exhuming them to take spores from the perithecia.

An interesting investigation can be carried out by placing threads of various cloths, like cotton, linen, terylene, etc., in flasks containing a culture medium made from:

Potassium dihydrogen orthophosphate (KH_2PO_4)	2.5 g
Ammonium sulphate (($NH_4)_2SO_4$)	2.0 g
Magnesium sulphate ($MgSO_4.7H_2O$)	2.0 g
Potassium chloride (KCl)	0.5 g
Calcium chloride ($CaCl_2$)	0.1 g
Thiamin	0.002 g
Water	1 litre

together with spores of *Chaetomium*. After about a month's incubation at room temperature, the threads should be compared with controls for tensile strength by loading with weights to breaking point.

TEACHING POTENTIAL

As in the previous chapter on bacteria, only a few of the vast number of possible investigations with fungi have been mentioned here. The dung of herbivorous animals provides a rich source of coprophilous fungi which can be easily seen if samples are kept moist. Rabbit droppings are perhaps the most rewarding for the teaching laboratory as well as being the least offensive to handle. Good growths can be obtained by placing them on a layer of plain water agar (1.5% agar) in a Petri dish. Faeces

from larger animals need to be kept on damp sterile sand under glass bell jars. The Butt Rot of conifers caused by *Fomes annosus* spreads from stumps left standing after felling. Where it is not possible or economic to lift and burn stumps they can be treated chemically to kill fungi but a more efficient procedure is to inoculate stumps with the non-pathogenic fungus *Peniophora gigantea* which is a superior competitor. Quite a range of laboratory investigations could be designed with these two fungi and coniferous wood. Is *Peniophora* equally effective at inhibiting *Fomes* in say pine, larch and spruce? Does it matter in terms of growth success how quickly *Peniophora* is applied to a cut surface of wood already carrying *Fomes*?[34]

But these problems—and others like them—are the real material of biological education and have been left undiscussed in this chapter so as to provide starting points for open-ended projects. The investigations given vary in complexity, they require differing levels of skill but all of them have been used successfully under the conditions of the teaching laboratory.

Chapter 4 Little plants and animals

Any body of water, be it fresh or salt, is rapidly colonised by small plants and animals.[57,60] Many of these early colonisers are microscopic and a host of them have bodies of one cell only. These algae and protozoa often occur in tremendous numbers and massive aggregations of particular species can colour bodies of water in quite spectacular ways. Further the individual forms of these organisms are often themselves extraordinarily complex and extremely beautiful to observe with the aid of the microscope.

All the basic skills of biological science, and a few additional ones, are needed to work in the field of microbiology. Students coming fresh to the study of micro-organisms need to build up experience in the field of (1) morphological studies which can lead to the segregation of species with a full appreciation of the variability in form which occurs in any species, (2) physiological investigations of external nutritive needs of organisms and of their internal metabolism and reproduction, and (3) their place in ecosystems.

Bacteria and fungi can be used to give this kind of experience as in the case of the Winogradsky Column (Chapter 1, page 6), the selective culture of bacteria (Chapter 2, page 30) and the nutrient needs of fungi (Chapter 3, page 47). But the small size of bacteria and the coenocytic nature of many fungi do make morphological studies particularly difficult so that the slightly larger bodies of the unicellular algae and the protozoa have real teaching advantages. Though some of the larger water plants like the Charophytes (e.g. *Chara, Nitella, Tolypella*) provide ideal material for the examination of plant cell structure since, for example, the internodal cells of *Chara* are anything up to 3.0 cm in length and furthermore cytoplasmic streaming can invariably be demonstrated. Unfortunately many species of *Chara* possess cortical cells which surround and mask the central cells and they may also contain deposits of calcium carbonate which hinder observations. None of the species of *Nitella* have these disadvantages and so where they are available they are to be preferred.

The unicellular, motile algae are ideal organisms with which to demonstrate the phenomenon of

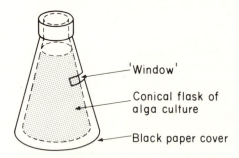

Figure 25 Blacked-out flask for Euglena

phototropism. *Euglena* spp. or *Chlamydomonas* spp. are probably the most suitable to use in simple culture in a conical flask which has been suitably 'blacked-out' with paper or paint, apart from a small aperture (see Fig. 25). This has merely to be left in a window with the aperture pointing towards the light for the algae to be seen to have congregated around the aperture. This simple investigation can be performed with differing species to illustrate the range of motility of algal cells and also the effect of various nutrient media on the motility (e.g. it is interesting to explore the importance of an ample nitrogen source in the medium).

A DEMONSTRATION OF POPULATION GROWTH [AL]

1 Preparation of population

The unicellular alga *Chlorella* can be conveniently used to demonstrate changes in population number with time if cultured in a cool room in the following medium.

0.5M Potassium nitrate (KNO_3)	30 cm^3	(50.5g/l)
0.5 M Dipotassium hydrogen orthophosphate (K_2HPO_4)	12 cm^3	(82 g/l)
0.5 M Magnesium sulphate ($MgSO_4$)	12 cm^3	(124 g/l)
0.5 M Iron (II) sulphate ($FeSO_4.7H_2O$)	6 cm^3	(140 g/l)
Distilled water	90 cm^3	

54

Figure 26 An aerated algal culture

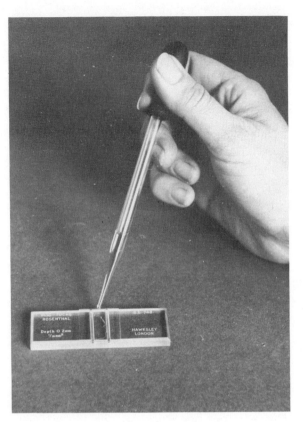

Figure 27 Loading a haemocytometer slide

The 150 cm^3 of medium should be placed in a 250 cm^3 conical flask and inoculated with the alga. It is best to avoid direct sunlight but diffuse light, as from a north facing window, and aeration—say from an aquarium pump—is essential. The numerical data is simply obtained by withdrawing aliquots daily and counting the cell numbers on a haemocytometer slide (see Chapter 2, page 34) (Figs. 26 and 27).

2 Examining the data

Under temperate conditions there is normally a lag phase of about 24 hours for *Chlorella* after which the medium suggested will support a log phase of growth for another nine days as shown in Figure 28.

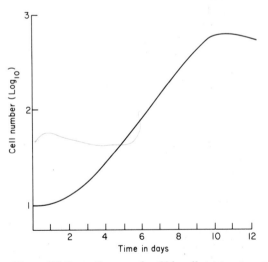

Figure 28 Growth curve for Chlorella at temperate room temperature

Experimental cultures of *Chlorella* are simple to set up and it is fairly easy to arrange to count the cells in a culture once a day over a period of 7—10 days. If students plot the actual number of cells observed against time, they will obtain a curve and the question naturally arises as to whether this is logarithmic or not. This question is simply tested by plotting the logarithm of the growth figures against time and establishing whether the result is a straight line—or nearly so! Students rapidly realise that looking up logarithms can be a rather tedious business and so they can be introduced to 'log-normal' graph paper at a stage where they will value the ease of calibrating this arithmetically along the horizontal axis and logarithmically up the vertical axis.

INVESTIGATING THE GROWTH REQUIREMENTS OF TWO ALGAE [AL]

1 Preparing nutrients

A range of nutrient deficient media and a complete medium for algal growth can be prepared from the chemicals shown in Tables 15 and 16. When making up these solutions, it is important to use only 'Analar' grade chemicals. The final solutions or modified forms with agar should be sterilised at approximately $103.5 kN/m^2$ for 15 to 20 minutes before use.

2 An investigation using Hydrodictyon

Take a stock of *Hydrodictyon* and introduce six specimens into a series of eleven $500 cm^3$ flasks each containing $250 cm^3$ of one of the nutrient solutions 1—11 (Table 16). With care it is possible to choose specimens of a similar size from the stock culture. Bung the experimental flasks with cotton wool and keep them in a cool room or a refrigerator set to $17°C$ both having a light source. Aeration is unnecessary but the flasks should be shaken occasionally. The results can be observed after one or two weeks.

3 An investigation using Chlorella

Prepare a suspension of *Chlorella* cells by washing the contents of a slope culture into a flask with distilled water and make it up to $250 cm^3$. Using a graduated pipette and taking care to shake the flask continuously, introduce two or three $0.1 cm^3$ samples of the suspension on to a range of algal nutrient agar surfaces as shown in Figure 29. These

Nutrient Source Solution	Chemical	Percentage concentration by weight
A	Calcium nitrate, $Ca(NO_3)_2.4H_2O$	7.9
B	Potassium nitrate, KNO_3	3.4
C	Sodium nitrate, $NaNO_3$	2.8
D	Potassium sulphate, K_2SO_4	2.9
E	Sodium sulphate, $Na_2SO_4.10H_2O$	2.2
F	Calcium sulphate, $CaSO_4.2H_2O$	0.17
G	Magnesium sulphate, $MgSO_4.7H_2O$	3.7
H	Sodium dihydrogen orthophosphate, $NaH_2PO_4.2H_2O$	2.1

Nutrient Source Solution	Chemical		Percentage concentration by weight
J	Magnesium nitrate, $Mg(NO_3)_2.6H_2O$		3.8
L	Micro-nutrient Stock Solution Manganous chloride, $MnCl_2.4H_2O$	2.0g	
	Copper chloride, $CuCl_2.2H_2O$	0.2g	
	Zinc chloride, $ZnCl_2$	0.3g	
	Boric acid, H_3BO_3	2.0g	
	Sodium molybdate, Na_2MoO_4	0.05g	
M	Boric acid, H_3BO_3		0.02
N	Micro-nutrient Stock Solution but with the Boric acid (H_3BO_3) omitted		

Table 15 Chemicals to prepare algal nutrient solutions

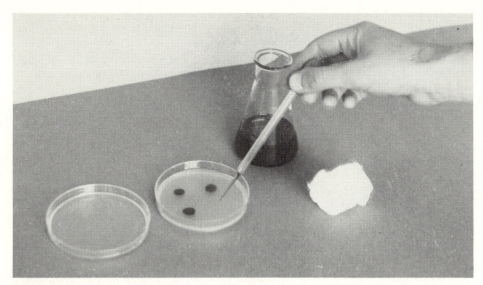

Figure 29 Inoculating Chlorella onto an agar plate

surfaces are prepared by setting up a series of Petri dishes each containing 15 cm^3 of one of the nutrient solutions (1−11, Table 16) plus 1.5% of agar. The inoculated plates should be incubated under the same conditions as those given for *Hydrodictyon* and after one to two weeks the growth from the inoculation sites can be measured by area or by taking two diameters at right angles.

PRODUCTION OF GROWTH INHIBITORS BY ALGAE

1 An antibiotic from Chlorella

Some plants produce chemical compounds which inhibit the growth of others which might compete with them. The micro-organism *Chlorella* produces a substance 'chlorellin' which is able to inhibit other micro-organisms around it. Thus, if we define an

	Nutrient Solutions	A	B	C	D	E	F	G	H	J	K	L	M	N
		cm³ of nutrient source solutions												
		to be mixed and made up to 500 cm³ with distilled water												
1	Complete nutrients	5	5	—	—	—	—	5	5	—	5	½	—	—
2	Phosphorus deficient	5	5	—	—	5	—	5	—	—	5	½	—	—
3	Potassium deficient	5	—	5	—	5	—	5	5	—	5	½	—	—
4	Calcium deficient	—	5	10	—	—	—	5	5	—	5	½	—	—
5	Nitrogen deficient	—	—	—	5	—	115	5	5	—	5	½	—	—
6	Magnesium deficient	5	5	—	—	10	—	—	5	—	5	½	—	—
7	Sulphur deficient	5	5	—	—	—	—	—	5	5	5	½	—	—
8	Iron deficient	5	5	—	—	—	—	5	5	—	—	½	—	—
9	Boron deficient	5	5	—	—	—	—	5	5	—	5	—	—	½
10	Micronutrient deficient (except boron)	5	5	—	—	—	—	5	5	—	5	—	½	—
11	Control (plain water)	—	—	—	—	—	—	—	—	—	—	—	—	—

Table 16 Recipes for mixing various nutrient source solutions to give experimental nutrient solutions

antibiotic as 'a substance produced by one micro-organism suppressing the growth of another' then 'chlorellin' which has been shown to be a breakdown product of algal fatty acids, is an antibiotic.

Laboratory demonstration of the production of 'chlorellin'.

The breakdown of the chlorellal fatty acids to make 'chlorellin' only occurs in the light and can be demonstrated simply in the teaching laboratory as follows:

1 prepare a medium in distilled water of the following composition:

    Proteose peptone   0.5%

    Agar powder   1.5% and sterilise it at approximately $103.5 kN/m^2$ for 10 minutes

2 place 15 cm^3 aliquots of this into sterile Petri dishes and allow it to set.

3 streak on a *Chlorella* suspension and incubate the plates at cool room temperature for three days, in as bright a light as possible (beware overheating).

4 streak across this a sample of 24 hour nutrient broth culture of *Staphylococcus albus* and leave for two days at room temperature.

5 observe and consider any zones of inhibition.

2 Antibiotics from larger algae

Some of the larger marine algae also produce antibiotic type substances. This can be readily demonstrated by using the red alga *Polysiphonia lanosa* which is epiphytic on the intertidal brown fucoid alga, *Ascophyllum nodosum*. This habit means that *Polysiphonia* is easily found and identified by a beginner.

To demonstrate the production of antibiotic by *Polysiphonia*:

1 detach the red alga from its host and rinse off excess sandy material.

2 surface sterilise it in dilute bleach (10% sodium hypochlorite solution) following by rinsing in sterile distilled water.

3 make a medium of the following composition:

    Dextrose   15 g

    Peptone   10 g

    Agar   15 g

    Water to one litre

4 dispense medium into universal bottles in 15 cm^3 aliquots and autoclave it at approximately $103.5 kN/m^2$ for 15 minutes.

5 introduce 0.1 cm^3 of a nutrient broth culture (24 hours at $37^{\circ}C$) of *Staphylococcus albus* into a sterile Petri dish.

6 add to the broth a cooled, melted aliquot of medium and mix thoroughly by gentle swirling.

7 before the agar has 'set' introduce a sample of the surface-sterilized seaweed.

8 when 'set' invert and incubate at $37^{\circ}C$ for 18 hours.

9 observe and consider any zones of inhibition.

PREPARING A PURE STRAIN CULTURE BY CLEANING

Methods

Axenic or pure strain cultivation of unicellular organisms like algae or protozoa is never simple and in the case of many phagotrophs it may well be impossible. However, many phototrophic protista may be grown axenically in relatively simple media if they can first be freed from contaminating bacteria. Several methods are commonly employed for cleaning organisms like serial washing, differential centrifugation, antibiotic treatment and migration through liquid or across solid media.[55,61,62] In any particular case, the method chosen will depend upon the organism to be decontaminated and it is usually necessary to employ a combination of two or more methods.

An example of possible procedures

The flagellate *Euglena gracilis* in common with most photosynthetic organisms, shows pronounced positive phototropism. This characteristic makes it amenable to cleaning by migrational methods.

1 Take a U-shaped tube filled with a sterile *Euglena* medium made from:

    Sodium acetate ($CH_3.COONa.3H_2O$)   0.1 g

    Beef extract   0.1 g

    Yeast extract   0.2 g

    'Bacto' tryptone   0.2 g

    Calcium chloride ($CaCl_2$)   0.001 g

    in 100 cm^3 water

Mask one side of the U-tube with black paper or polythene as in Figure 30. Place a pipette full of thick crude contaminated *Euglena* culture in the darkened end of the tube, plug the ends of the tube lightly with cotton wool and illuminate the uncovered end for two hours. After this time a few of the *Euglena* should have reached the lit end of the tube. Remove the light and the plug, and using a sterile pipette, remove some of the liquid from the lit end. Use this liquid to inoculate about ten tubes of *Euglena* medium at the rate of one drop per tube.

Essentially each tube should contain about six *Euglena*.

Finally, incubate these tubes beneath warm white fluorescent lamps for about one week and examine them for growth of *Euglena*. It may be necessary to incubate for longer than one week because of a very small inoculum. If growth of *Euglena* has taken place, check for contamination by inoculating a loopful of culture on to a nutrient agar slope and examine this slope for bacterial growth after a few days.

2 *Euglena gracilis* can be cultured on agar plates and single individuals will give rise to compact colonies. It is theoretically possible, therefore, to obtain clean *Euglena* colonies by normal plating methods. However, microscopical examination of a normal *Euglena* culture will reveal that bacteria outnumber *Euglena* by hundreds to one. The chance of obtaining discrete colonies by direct plating is therefore very remote. Bacterial numbers must first be drastically reduced. This can be done by centrifugation.

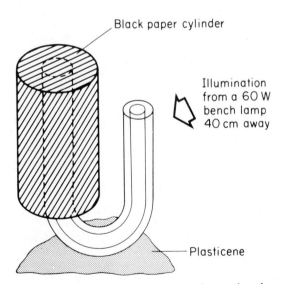

Figure 30 A masked U-tube for Euglena migration

Put a little *Euglena* culture into a centrifuge tube and spin gently in a bench centrifuge for one minute. A firm pellet of *Euglena* should result. Pour off the supernatant and resuspend the organisms in sterile medium. Repeat this procedure eight times. After pouring off the supernatant for the eighth time, shake the tube gently to suspend the *Euglena* in the small amount of liquid remaining. Flame a loop, allow it to cool and streak *Euglena* on to *Euglena gracilis* medium agar plates, which have been prepared by adding 1.5 g of agar to the liquid medium. Inoculate four plates. Examine some of remaining *Euglena* from the tube and also some of the original culture by phase contrast. Make a quick estimation of the *Euglena* bacteria ratio in each case and try to assess the 'state of health' of the *Euglena* after centrifugation.

Incubate the plates beneath fluorescent lamps and examine inverted plates beneath a binocular microscope without removing the lids, before one week has elapsed if possible. If discrete colonies of *Euglena* have developed, pick these off with a sterile micropipette or loop and inoculate on to fresh plates or into *Euglena* medium. Check subcultures after one further week for signs of *Euglena* growth and bacterial contamination.

MAINTAINING AXENIC CULTURES

Once such cultures of algae or protozoa have been prepared or purchased, they should be maintained on soft proteose peptone agar slopes prepared from:

Proteose peptone (Oxoid)	1.0 g
Potassium nitrate (KNO_3)	0.2 g
Dipotassium hydrogen orthophosphate (K_2HPO_4)	0.02 g
Magnesium sulphate ($MgSO_4.7H_2O$)	0.02 g
Agar	10.0 g

Distilled water to 1 litre

Important points to note are: (1) the agar (1%) is soft (dilute); (2) that it should be laid in the slope in as thick a layer as possible, the culture tube should be filled almost to the plug; (3) before inoculation, the slope should be squirted with sterile water from a Pasteur pipette, so as to leave perhaps 0.5 cm^3 or less standing in the bottom of the tube; (4) good firm plugs are essential to success; (5) it is advantageous to cover the plug with an ample cover of metal foil. For

59

non-photosynthetic species store the tubes upright in a beaker or glass jar, the floor of which is protected with cotton wool. The container itself is then covered with a generous cap of metal foil. These precautions ensure minimal evaporation and maximal protection from dust, which is helpful to the success of the next transfer.

MAKING COTTON WOOL PLUGS

In order to store cultures of algae and some protozoa, it is important to use firm plugs for flasks and tubes. It is worth the trouble of using the following procedure to prepare them:

1   take a sheet of good brown wrapping paper, turn it rough side up and stick it to the bench with self adhesive tape.
2   for tubes tear off about 20 cm of cotton wool from the roll; for flasks, 30 cm. For boiling tubes and flasks use the full thickness of cotton wool; for test tubes use 1/2 to 2/3 of the full thickness.
3   put the piece of cotton wool on the brown paper and roll it with the fingers and thumbs, *away* from you, until the roll reaches the desired thickness.
4   cut to the desired length with a sharp razor-blade.
5   roll once more with the palm to correct thickness, and trim off the surplus with scissors.

When a variety of cultures are kept in store, it is easier to distinguish algae if they are plugged with green non-absorbent cotton wool.

THE ISOLATION AND CULTURE OF SMALL AMOEBAE

In addition to using amoebae from laboratory cultures, it is interesting to attempt to obtain specimens from natural sources. Most small amoebae grow well on non-nutrient agar plates streaked with bacteria. *Escherichia coli* and *Klebsiella aerogenes* are particularly suitable for this purpose. The isolation and culture of small amoebae is therefore relatively simple.

Procedure

Liberally streak several plates of amoeba saline agar with *Escherichia coli* or *Klebsiella aerogenes* and place a small lump of worked soil, some activated sludge or some pond mud in the centre of each plate. Examine the plates after one week for growth of amoebae.

Amoebae will migrate across the plate in search of

food and upon encountering a patch of bacteria will feed and multiply. When a patch of bacteria has been exhausted, either encystment or further migration will take place. In this way, a gradual sorting takes place as the amoebae move across the plate and colonies of amoebae near the perimeter of the plate probably consist of individuals of the same species.

When plates are examined the observer will probably see many amoebae of assorted sizes. However, it should be possible to find colonies near the perimeter of the plate which contain amoebae all of similar appearance. Using a loop, transfer colonies to fresh plates previously streaked with *E. coli*.

Now examine some of the amoebae from the original culture by phase contrast. Do this by placing a few drops of Chalkley's Medium (see Chapter 1, page 16) on the plate, and suspend the amoebae in this liquid by scraping the plate with a loop. Take up the amoeba suspension with a pipette and transfer to a slide.

Examine subcultures for growth after one further week.

Recipe for amoeba saline agar

Agar	15 g
Calcium chloride (CaCl$_2$)	0.06 g
Sodium chloride (NaCl)	1.0 g
Potassium chloride (KCl)	0.04 g
Distilled water	1 litre

TEACHING METHOD

Sadly the old pattern of biology teaching required the study of certain protozoan and algal types and much of this rigidity of approach continues to-day. However, the range of non-standard types, of reasonable size, which occur in the environment is large, giving students ample opportunity to explore biological problems in a truly open-ended way. The possibilities are about as numerous as the grains of sand on a seashore but some successful starting references are given in Appendix 2.3.

As we have come to appreciate the value of biological, as opposed to zoological or botanical, education in the twentieth century, so we have realised a need to give more time to the field of microbiology. Certainly this field of study produces an ample crop of stimulating surprises at all levels but it does provide one of the most cogent links between plant and animal sciences.

Appendix 1 Sources of information, publications, sources of apparatus, chemicals and living materials

1.1 Information
1.2 Publications
1.3 General Biological Suppliers
1.4 Other Suppliers of Apparatus and Chemicals
1.5 Living Materials

This list is not completely comprehensive, nor does the inclusion of an address imply that any particular publication, supplier or piece of apparatus is to be preferred to any other. Before buying it is generally advisable to consult the catalogues of more than one supplier.

In order to avoid repetition the address of each supplier is given in full on the first occasion of mention only. All subsequent references are by name only and with the sub-section number in which the full address may be found. The list is annotated where necessary.

Every care has been taken in compiling this list; however no responsibility can be accepted for any inaccuracies.

It is suggested that teachers, rather than individual pupils should apply for information.

1.1 INFORMATION
Microbiology in Schools Advisory Committee (MISAC), Hon. Secretary: Dr. G Holt, Department of Life Sciences, Polytechnic of Central London, 115 New Cavendish Street, London W1M 8JS

Public Health Laboratory Service, Central Public Health Laboratory, Colindale Avenue, London NW9 5HT

1.2 PUBLICATIONS
'American Biology Teacher' National Association of Biology Teachers, 1420 N. Street, N.W., Washington D.C. 20005, USA

H M S O Publications, Government Bookshop, P O Box 569, London SE1 9NH

'Journal of Biological Education' Institute of Biology, 41 Queens Gate, London SW7 5HV

'Natural Science in Schools' School Natural Science Society, General Secretary: Miss M J Sellers, 2 Bramley Mansions, Berrylands Road, Surbiton, Surrey KT5 8QV
Publications Officer:— 44 Claremont Gardens, Upminster, Essex RM14 1DN

'School Science Review' Association for Science Education, College Lane, Hatfield, Hertfordshire AL10 9AA

1.3 GENERAL BIOLOGICAL SUPPLIERS
These firms supply a wide range of living material as well as apparatus and chemicals.
Bioserv Ltd., 38 Station Road, Worthing, Sussex BN11 1JP

G B I (Labs) Ltd., Heaton Street, Denton, Manchester M34 3RG

T Gerrard and Co., Gerrard House, Worthing Road, East Preston, West Sussex BN16 1AS

Griffin Biological Laboratories Ltd., Gerrard House, Worthing Road, East Preston, West Sussex BN16 1AS

Philip Harris Biological Ltd., Oldmixon, Weston-super-Mare, Somerset BS24 9BJ

Timstar Biological Suppliers, Lower House, Little Budworth, Tarporley, Cheshire CW6 9BL. Mainly living organisms.

1.4 OTHER SUPPLIERS OF APPARATUS AND CHEMICALS
Astell Laboratory Service Co., 172 Brownhill Road, Catford, London SE6 2DL. Culture media.

Difco Laboratories, P O Box 14B, Central Avenue, Molesey, Surrey KT8 0SE. Culture media and reagents, biological stains.

Dyos Plastics Ltd., 242 Tolworth Rise South, Surbiton, Surrey KT5 9NB. Plastic disposable Petri dishes.

Esco (Rubber) Ltd., 14–16 Great Portland Street, London W1N 5AB. Rubber and plastic tubing, bungs; disposable plastic Petri dishes.

Flow Laboratories Ltd., Victoria Park, Heatherhouse Road, Irvine, Ayrshire KA12 8NB. Tissue cultures; tissue culture media and sera, viral and immunological reagents, microtitration equipment, tissue culture plastics and laboratory equipment.

Gallenkamp, A. & Co. Ltd., Technico House, Christopher Street, London EC2P 2ER. 'Parafilm' sealing tissue.

Halsey's Electrical Wholesale Co. Ltd., Brandon House, Wyfold Road, London SW6 6SQ. Ultraviolet lamps.

Hanovia Ltd., 480 Bath Road, Slough, Buckinghamshire SL1 6BL. Ultraviolet lamps.

Harris Philip & Co. Ltd., Lynn Lane, Shenstone, Staffordshire WS14 0EE. Foam plastic bungs.

Horwell, Arnold R., 2 Grangeway, Kilburn High Road, London NW6 2BP. Haemocytometers, disposable syringes.

Luckham Ltd., Labro Works, Victoria Gardens, Burgess Hill, Sussex RH15 9QN. Aluminium bottle racks; disposable plastic tubes and containers.

Microbiological Supplies, P O Box 10, Tunbridge Wells, Kent TN1 1SZ. Culture media.

Millipore (UK) Ltd., Millipore House, Abbey Road, Park Royal, London NW10 7SP. Filters.

Northern Media Supply Ltd., Crosslands Lane, Newport Road, North Cave, Brough, E. Yorkshire HU15 2PG. Plastic apparatus, autoclaves, incubators, water baths, culture media.

Oxoid Ltd., Wade Road, Basingstoke, Hampshire RG24 0PW. Culture media, Ringer's solution tablets and accessories for microbiology.

Payne, C.E. & Sons Ltd., 6 Iveley Road, London SW4 0HS. 'Cling' plastic sealing film.

Searle Scientific Services (G T Gurr), Coronation Road, Cressex Industrial Estate, High Wycombe, Buckinghamshire HP12 3TA. Biological stains.

Sterelin Ltd., 12–14 Hill Rise, Richmond, Surrey TW10 6UD. Plastic Petri dishes and autoclavable disposal bags.

1.5 LIVING MATERIALS
Living material may be obtained from the following sources in addition to the General Biological Suppliers 1.3.

Building Research Establishment, Princes Risborough Laboratory, Princes Risborough, Aylesbury, Buckinghamshire HP17 9PX. (Formerly the Forest Products Research Laboratory.) Fungi—wood rotting.

Commonwealth Mycological Institute, Collection of Fungus Cultures, Ferry Lane, Kew, Surrey TW9 3AF. Fungi (other than animal pathogens, wood rotting species and most yeasts). A leaflet listing those fungi suitable for teaching purposes is available.

Culture Collection of Algae and Protozoa, 36 Storey's Way, Cambridge CB3 0DT. Algae and protozoa (non-pathogenic).

Microbiological Supplies (1.4)

National Collection of Dairy Organisms, National Institute for Research in Dairying, Shinfield, Reading, Berkshire RG2 9AT. Bacteria from milk and milk products.

National Collection of Industrial Bacteria, Torry Research Station, P O Box 31, 135 Abbey Road, Aberdeen AB9 8DG. Bacteria of industrial and scientific interest.

National Collection of Marine Bacteria, Torry Research Station, P O Box 31, 135 Abbey Road, Aberdeen AB9 8DG. Bacteria of marine origin; organisms pathogenic for poikilothermic animals; extreme halophiles and luminous bacteria.

National Collection of Plant Pathogenic Bacteria, Ministry of Agriculture, Fisheries and Food, Plant Pathology Laboratory, Hatching Green, Harpenden, Hertfordshire AL5 2BD. Bacteria pathogenic to plants.

National Collection of Yeast Cultures, Brewing Industry Research Foundation, Lyttel Hall, Nutfield, Nr Redhill, Surrey RH1 4HY. Yeasts (other than pathogens).

Oxoid Ltd. (1.4) Discs impregnated with bacterial cultures, discs impregnated with antibiotics, School Microbiology Kit.

Appendix 2 Bibliography and References

SECTION 1   GENERAL   (1–26)

1   *American Biology Teacher* (1968) Microbiology 30 (6) Entire Issue.
2   Bainbridge B W (1971) Workshop: Micro-organisms in School Biology *Journal of Biological Education* 5 (6) 309–12.
3   Bainbridge B W (1972) Microbiology in Schools Advisory Committee (MISAC) *Journal of Biological Education* 6 (3) 207–10.
4   Bradbury S Rev. (1973) *Peacock's Elementary Microtechnique* 4th Edn. Arnold.
5   Broda P M A (1957) Some Experiments with Tobacco Mosaic Virus *School Science Review* 39 (137) 78–82.
6   Clowes R C and Hayes W Eds (1968) *Experiments in Microbial Genetics* Blackwell.
7   Collins C H and Lyne P M (1970) *Microbiological Methods* 3rd Edn. Butterworths.
8   Commonwealth Mycological Institute (1968) *Plant Pathologist's Pocketbook*.
9   Department of Education and Science (1976) Safety Series No. 2 *Safety in Science Laboratories*. Rev. HMSO.
10 Educational Use of Living Organisms Project (Schools Council) Comber L C (1976) *Organisms for Genetics* Hodder and Stoughton Educational.
11 Educational Use of Living Organisms Project (Schools Council) Kelly P J and Wray J D Eds (1975) *The Educational Use of Living Organisms. A Source Book* Hodder and Stoughton Educational.
12 Garbutt J W and Bartlett A J (1972) *Experimental Biology with Micro-organisms* Teachers Guide and Students Manual. Butterworths.
13 Holt G (1974) Some Tips on the Safe Handling of Micro-organisms in Schools *School Science Review* 56 (195) 248–252.
14 Holt G (1975) Local Advisers for the Teaching of Microbiology in Schools *Journal of Biological Education* 9 (1) 26–28.
15 Kelly P J and Wray J D (1971) The Educational Use of Living Organisms *Journal of Biological Education* 5 (5) 213–318.
16 Kendall M (1969) Preparation and Preservation Techniques *Housecraft* April.
17 Millipore (1969) *Experiments in Microbiology* Published by Millipore Corporation.
18 MISAC (1971) Review of Practical Manuals of Value to School Teachers Using Micro-organisms. *Journal of Biological Education* 5 (6) 331–40.
19 National Physical Laboratory (1972) *Changing to the Metric System* 4th Edn. HMSO.
20 Oxoid (1974) *Oxoid Manual* 4th Edn. Published by Oxoid Ltd.
21 Schatz A Brandon G G and Webber J D (1970) Sterilisation of Plastic Petri Dishes *American Biology Teacher* 32 (5) 294–5.
22 Schools Council (1974) *Recommended Practice for Schools relating to the Use of Living Organisms and Material of Living Origin.* Hodder and Stoughton Educational.
23 Shapton D A and Board R G (Eds) (1972) *Safety in Microbiology* Academic Press.
24 Sirokin G and Cullimore S (1969) *Practical Microbiology* McGraw-Hill.
25 Whittaker R H (1959) On the Broad Classification of Organisms. *Quarterly Review of Biology* 34 210.
26 Wyatt H V (1968) Teaching of Microbiology in Schools *Journal of Biological Education* 2 (3) 187–95.

SECTION 2. BACTERIA AND FUNGI (27–52)
27 Barnett H L and Hunter B (1972) *Illustrated Genera of Imperfect Fungi* 3rd Edn. Rev. Burgess.
28 Boothroyd C W and Kelman A (1966) Laboratory experiments in plant pathology. *American Biology Teacher* 28 (6) 478–502.
29 Bottle R T (1967) *Education in Chemistry* 4 197.
30 Cunnell G J (1963–1965) Studying Fungi in Schools *School Science Review* 45 (155) 93–102 and (156) 312–23; 46 (160) 579–91.
31 Dade H A and Gunnell J (1969) *Class Work with Fungi* 2nd Edn. Commonwealth Mycological Institute.

32 Finch I (1963) Experiments on Bacteria, Fungi and Decay *Natural Science in Schools* 1 (2) 38—47.

33 Fincham J R S and Day P R (1967) *Fungal Genetics* Blackwell.

34 Forestry Commission (1957) *Fomes annosus: A Fungus causing Butt Rot and Death of Conifers.* HMSO.

35 Freeland P W (1973) Some Practical Aspects of Sugar Fermentation of Baker's Yeast *(Saccharomyces cerevisiae) Journal of Biological Education* 7 (5) 14—22.

36 Garatun—Tjeldsto O (1970) Incorporation of Isotopes in Growing Cells of *Escherichia coli. Journal of Biological Education* 4 (2) 97—104.

37 Gilman J C (1957) *A Manual of Soil Fungi* Iowa State University Press.

38 Griwell A R and Jackson J F (1969) Microbial Culture Preservation with Silica Gel *Journal of General Microbiology* 58 423—25.

39 Harris P J (1969) A Demonstration of Microbial Antagonism in Soil Using *Serratia marcescens* as a Marker Organism. *Journal of Biological Education* 3 (3) 237—9.

40 Ingold C T (1961) *The Biology of Fungi* Hutchinson.

41 Jack K M Lapage S P and Shewan J M (1969) Supply of Bacteriological Cultures to Schools. *Education in Science* No. 3 29—30.

42 Lilly V G and Barnett H L (1951) *Physiology of the Fungi* McGraw-Hill.

43 McGrady M H (1918) *Canadian Public Health Journal* 9 201.

44 The Milk (Special Designation) Regulations (1971) Statutory Instruments No. 1571 HMSO.

45 Mullenger L and Gill Nijole R (1973) *Vibrio natriegens:* a Rapidly Growing Micro-organism Ideally Suited for Class Experiments *Journal of Biological Education* 7 (5) 33—9.

46 Perkins D D (1962) Preservation of *Neurospora* Stock Cultures with Anhydrous Silica Gel *Canadian Journal of Microbiology* 8 591—94.

47 Reports on Public Health and Medical Subjects (1969) No. 11 *The Bacteriological Examination of Water Supplies.* HMSO.

48 Schopfer W H (1945) Les Tests Microbiologiques pour la Determination des Vitamines *Experientia* 1 1—68.

49 Seaman A (1963) *Bacteriology for Dairy Students* Cleaver Hume.

50 Smith G (1969) *An Introduction to Industrial Mycology* E Arnold.

51 Swarcop S (1938) *Indian Journal of Medical Research* 26 353.

52 Swarcop S (1951) *Indian Journal of Medical Research* 39 107.

SECTION 3. ALGAE AND PROTOZOA. (53—66)

53 Andresen N (1956) Cytological Investigations in the Giant Amoeba *Chaos chaos* C R Laboratory Carlsberg 29 435—555.

54 Brown T J (1967) Comparative Ecology of Protozoa. A Laboratory Exercise for Undergraduates *Journal of Biological Education* 1 (2) 161—7.

55 Brunel J et al Eds (1950) *The Culturing of Algae* Charles F. Kettering Foundation.

56 Goulding K H and Merrett M J (1970) Experiments on Enzyme Induction in *Chlorella pyrenoidosa Journal of Biological Education* 4 (1) 43—52.

57 Jahn T L and Jahn F F (1949) *How to Know the Protozoa* W C Brown.

58 Mackinnon D L and Hawes R S J (1961) *An Introduction to the Study of the Protozoa* Oxford.

59 Needham J G et al (1959) *Culture Methods for Invertebrate Animals* Dover, New York. Originally published 1937 by Comstock Publishing Company.

60 Prescott G W (1954) *How to Know the Freshwater Algae* W C Brown.

61 Pringsheim E G (1945) Cultures of Algae and Flagellates *School Science Review* 26 (99) 198—207.

62 Pringsheim E G (1963) *Pure Cultures of Algae: Maintenance* Hafner, New York.

63 Sohn B I (1972) Algae as Pollution Indicators: Analysis Using the Membrane Filter *American Biology Teacher* 34 (1) Jan. 19—22.

64 Taylor M (1952 and 1957) Recent Advances in *Amoeba* lore Parts 1 and 2 *School Science Review* 34 (122) 98—108 and 38 (136) 384—90.

65 Vickerman K and Cox F E (1967) *The Protozoa* Murray.

66 West G S and Frisch F E (1927) *A Treatise on the British Freshwater Algae* Cambridge.